Y0-CXA-177

The Martin Murphy Family Saga

The route of the Stephens-Murphy Party, Council Bluffs, Iowa to California. FROM *TRAIL OF THE FIRST WAGONS OVER THE SIERRA NEVADA* BY CHARLES K. GRAYDON.

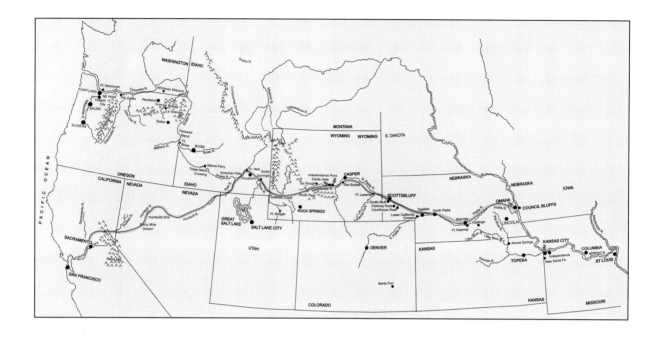

The Martin Murphy Family Saga

By Marjorie Pierce

Foreword by Dianne McKenna
Introduction by Mike Malone

Published by:
The California History Center & Foundation
Cupertino, California
Local History Studies Volume 38

Copyright © 2000.

Managing editor, N. Kathleen Peregrin
Editor, Lisa Christiansen

Published by the California History Center & Foundation,
De Anza College, 21250 Stevens Creek Blvd., Cupertino, CA 95014.

Library of Congress Cataloging-in-Publication Data

Pierce, Marjorie.
    The Martin Murphy famil saga / by Marjorie Pierce.
        p. cm. — (Local history studies ; v. 38)
    Includes bibliographical references (p.) and index.
    ISBN 0-935089-23-3 — ISBN 0-935089-24-1 (soft cover)
        1. Murphy, Martin, 1807-1884. 2. Murphy family. 3. Pioneers—California—Santa Clara
County—Bibilography. 4. Businessmen—California—Santa Clara County—Biography. 5.
Irish Americans—California—Santa Clara County—Biography. 6. Santa Clara County
(Calif.)—Biography.7. Santa Clara County (Calif.)—History—19th century. I. Title. II.

F868.S25 P63 2000
979.4'730049162—dc21
                                                                                00-044498

Disclaimer:

Direct quotations may not always meet the standards for exactness. However, the conversation
as presented is meant to enhance the truth of what is happening.

Funded through generous donations from:
John and Helen Cahill
Helen K. Cahill
Hugh Stuart Center Charitable Trust
Susan and George Fesus
Moira K. Holden
Emily and Ernest Renzel, Jr.

Cover Design by: Douglas E. Wood
Desktop Design by: Metro Design Center Inc.
Saratoga, CA (408) 253-0466

Acknowledgments

Introduction by Michael S. Malone

Forward by Dianne McKenna

Graphics: Long Pham

My Personal Computer Guru: Dan Meadows

So many friends who helped along the way reading the manuscript
include Earle Schmidt, Paul Lukes, Cecily Kyes, Anita Hawkes.

Lisa Christensen: Librarian — California History Center & Foundation

A special thanks to Mary Jo Ignoffo.

In loving memory of my mother: Margaret Mary Murphy Upstill

Contents

Foreword
by Dianne McKenna

Far into the night multitudes lingered among the illuminated groves,
for the scene was one of unsurpassing loveliness...
one that has never before or since been witnessed in California,
and one that will long be remembered - Hugh Bancroft

Too often Americans know more about their nation's history than they do about their local history. Marjorie Pierce is dedicated to changing that. The Martin Murphy Family Saga is her latest book on the pioneers who endured swollen rivers, buffalo stampedes, disease, impassable mountains and freezing snow to reach a land of perennial sunshine and fertile soil. The Murphy-Stephens overland party was the first to cross the summit into California with wagons and cattle. They proved that families and supplies could reach this country's western shore.

Unfortunately, what remains of the Murphy Family in Sunnyvale is held in the memory of too few residents. Marjorie Pierce's book will not only refresh and restore our human memory, but it can be added to all memory banks available today and quickly accessed by future generations. As we move to the new Millennium, it is important to understand what inner drive motivates today's pioneers, just as it is important to understand what motivated the Murphy family 150 years ago. This is a human lesson as much as it is an historical one.

In the last chapter of the book, Marjorie recounts the Golden Wedding Party. "In 1881 Martin Murphy, Jr. and Mary Bolger celebrated their 50th wedding anniversary on the grounds of their Bay View Ranch home. The press estimates that between 5,000 and 10,000 people arrive by train (chartered by the host), in carriages, buggies, farm wagons and on horseback. Captain Isaac Branham, California's most famous barbecuer, comes up from Los Angeles to preside over the

114-foot long barbecue pit, and does the double duty of serving as master of ceremonies. It is written that never before have so many prominent people in California assembled at one event—United States Senators, state legislators, multi-millionaires, black suited Jesuits from Santa Clara College, and black and white-garbed nuns from Notre Dame College, both of which Martin Murphy helped to found. The courts in San Francisco and San Jose adjourn for the day so the judges as well as the rest of their courts can attend.

In 1981, the City of Sunnyvale celebrated the centennial of the Murphy's 50th anniversary party. In historical tradition, it was called, "The grandest party the West has ever seen." The celebration began with a gala anniversary ball on Friday night and continued with an all-day party on Saturday, highlighted by a barbecue. In keeping with the spirit of the Murphy's party, married couples celebrating their 50th anniversaries were honored at the festivities, a special train was commissioned to bring people to the party, floral arrangements replicated the original designs, and public announcements were distributed, inviting people to attend. Not unlike the account of the original party, I'm sure press estimates would be that between 5,000 and 10,000 people attended. I know the Murphys would have enjoyed the party, even though it could not possible match the significance of the party they gave 100 years earlier.

The following year, 1982, our family took a trip to Ireland and made a point of visiting the hometown of the Murphy family in County Wexford. I wrote to the county officials that I wanted to bring greetings from Sunnyvale, since the city was founded by one of their native sons. A warm sense of welcome came over us as we drove up to the county building and saw both the Irish and American flags waving in the breeze. We couldn't have felt more welcome if we, ourselves, had been the Murphys.

During the same trip, I learned how time and history are related. An Irish man who lived on the country's western coast told me that his family originated from the county to the north and that the family had recently moved to the coast. I asked, "When did your family move to the coast?" expecting him to answer a year or two earlier. "We moved here in 1651," he responded. By his standards, the Murphys moved to Sunnyvale recently. He knew his family's and the town's histories—at least from 1651. How much richer our education would be if we, too, knew our town's and its people's history from the time of founding.

People like Marjorie Pierce are our connection to our history and our past. They also provide us a view of the future. Marjorie gives us a treasure to enjoy.

DIANNE McKENNA

Dianne McKenna is a former mayor of Sunnyvale and Santa Clara County supervisor and a well-known community activist and philanthropist.

Introduction
by Michael S. Malone.

You are about to meet a remarkable family, and most of all a remarkable man.

California, especially San Francisco is justly celebrated as the heartland of entrepreneurship. But, who was the greatest entrepreneur in California history? The obvious names quickly spring to mind: Hewlett and Packard, Robert Noyce of Fairchild and Intel, Steve Jobs of Apple, Jim Clark of Silicon Graphics and Netscape, perhaps a half-dozen more celebrity billionaires. One might add to the list, Leland Stanford, the Lockheed brothers and from Hollywood Jack Warner and George Lucas. All larger-than-life figures, who created vast empires seemingly from nothingness.

And yet, to my mind, all of them fade in comparison to Martin Murphy, Jr. When San Francisco was merely a tent-city, and Silicon Valley a vast meadow dotted with oak trees, Murphy literally began with nothing... and built one of the greatest personal empires in history. Converted to contemporary dollars, especially given the industrial parks now scattered over that landscape, Murphy's wealth would dwarf even that of the most successful contemporary cyber-tycoon.

Murphy was not only the prototype of the California tycoon (he predates Stanford and the rest of the railroad tycoons by a decade), but he towers over the "other" California—that is, not San Francisco or L.A.—during the 19th century. He is one of those extraordinary Victorian figures, who always seems to be at the right place when history occurs, whose experiences are almost impossibly heroic, and whose ultimate triumph is almost unimaginable in scale. There is the California Trail, Sutter's Fort, the discovery of gold; the Donner Party, some of whose members, after

coming down from the Lake were taken in by the Murphys; the Bear Flag Revolt following the first act of the Mexican War that occurred on the Murphy Ranch on the Cosumnes River. Then there are the stories, like the naming of baby Elizabeth Yuba, that seem more fantastic to me now than when I was a boy.

And that's just the beginning, a mere three years in a long life. Ahead lay thirty more years of immense power and wealth, of forging a new community—and ultimately a New World. The Murphy children whose story is also told in the biography were great consolidators. Just as their father was the entrepreneur, they were the senior executives of a new society, setting new rules and founding new institutions that govern us in the Valley even today. If Martin's is an adventure story, that of his progeny is a tale of manners. Just as Martin helped found Santa Clara University and College of Notre Dame, his son, Bernard, brought the State Normal School (San Jose State University) to San Jose. This is a subject Marjorie knows well, and her understanding shines on every page.

So welcome to the world of the Murphys, a tale so fabulous that no writer would dare invent it. And in this story, thanks to the craft of Marjorie Pierce, we modern Silicon Valleyites, can discern the first traces and patterns of ourselves.

MICHAEL MALONE

Michael S. Malone, a native of Sunnyvale, is a journalist, author of The Big Score: The Billion Dollar Story of Silicon Valley, The Virtual Corporation *and* The Microprocessor: A Biography *and the host of his own weekly public television program.*

CHAPTER ONE
Wexford: 1820

When Martin Murphy, a warm, generous man of 35 years, and his wife, quiet, dark-haired, blue-eyed Mary Foley, invited neighbors to their farm at Ferns in County Wexford, it was intended to be more than a social evening. They had arrived at a decision—one they had long pondered, and one they now wanted to share with their friends.

Martin finally asked for everyone's attention. With Mary standing beside him, his usually affable manner turned serious as he announced to those gathered, "We are leaving our beloved Ireland and moving our family to North America." He seemed not to notice their sighs of disappointment as he hurried on. "As you know, Mary's brother, Matthew, and his wife, Elizabeth, went to America last year and they are happy with the freedom they have found there.

"A new farming community is being developed in Canada and Mary and I feel it offers us and our children a chance to be free from the domination of the British. Ever since I took part in the Rebellion of 1798, even though I was but a lad, I have felt a shadow over my head.

"True," he continued, "the plight of the Catholics in Ireland has improved and many fine Protestants such as Parnell have espoused our cause, but still we do not have the right to vote or to hold public office." His choler rose as he said that, while the Protestants are absentee landlords and the ruling class, one-third of our people are starving. "We all believe the Protestant plan is to destroy Catholicism in Ireland and reduce us Catholics to total ignorance. Still we have to pay tithe to support the state church."

As he finished his spirited discourse, everyone started talking at once. No one disagreed with what Martin had been saying, but, as one farmer added, "Martin, my dear fellow, you didn't go far enough. You failed to mention the economy—that ever since Napoleon's downfall, the grain market has dropped almost out of sight. The landlords are putting the land into grazing and are evicting us tenants. We don't know who will be next."

Martin agreed. He was one of the larger, more successful farmers and so far he had been able to hold on.

Continuing with their plans he said they would leave 13-year-old Martin, Jr., a responsible lad who had worked with him on the farm, to take over until he reaches 21. "He and his sister, Margaret," he said, "will stay with my mother. Since my father and brother were lost at sea she has lived alone."

Mary, whose face showed her anxiety over leaving her children, interrupted her husband to say, "Martin and I have troubled over this decision for some time, but we have to be concerned about the future of our other four children. They must live in a free country where they can grow up, get an education and have the right to vote."

The Murphy crest. From Clans of Ireland *by John Grenham, The Wellfleet Press.*

1

Listening to the discussion Patrick Bolger, whose land adjoined the Murphys' farm, became increasingly thoughtful. Finally he said, "I also took part in the Rebellion and I have seen what the British have done to those who did. I agree with Martin that we are still suspect." It didn't take Pat long to decide he, too, would move to Canada with his family.

Oblivious to all this conversation, Patrick's lively twelve-year-old daughter, Mary, was joyfully dancing a jig with the serious Martin, Jr. They would not see each other again until their meeting eight years later in Quebec.

With all their worldly belongings packed in boxes and barrels and loaded on their wagon, Mary and Martin bade their eldest children, young Martin and Margaret, a tearful farewell. Joined by their four younger children, James, Mary, Bernard and Johanna, they set forth from Ferns. Looking back at the stone ruins of Ferns Castle, Martin reminded them not to forget their historic past—that Ferns was the Royal Seat of the Kings of Leinster, among the greatest historical centers in Ireland. As they passed the graveyard, Martin grew silent as he recalled the execution of Father Murphy and the burning of his headless body in a barrel of pitch—and of brave women, for example a sister of Father Murphy, who collected the remains and placed the bones in the tomb of the parish priest, Father Andrew Cassin.

When they reached the quaint little town of Enniscorthy on the banks of the River Slaney, Martin pointed up to Vinegar Hill, site of the first attack of the 1798 Rebellion, and where the rebels met their final defeat. He told them about Father Murphy, leader of the rebellion, and said that after hearing his impassioned speech, he signed up. One day, he predicted

(and as it turned out correctly), "a statue of brave Father Murphy, with pike in hand, will stand in the town square." Although Martin could not have known it then, one of his fellow soldiers in that rebellion was the great-grandfather of a man who would one day become president of the United States...John F. Kennedy.

On their arrival at Wexford harbor, Martin once more tried to impress upon his children their heritage because he was sure they would never again see their native land. "Wexford," he said, "was a port for invaders for over six hundred years. In this harbor the Norsemen invaded Ireland in the eleventh century. In the 1200's the Normans came."

Martin added another bit of Wexford pride as he predicted, correctly, "One day there will be a statue in the town square of a heroic young man from Wexford named John Barry. He was an admiral in the American War of Independence against the British. They called him the "Father of the American Navy."

The children, by this time, had had enough history. Led by their older brother, Jimmy, the excited children set out to find the sailing ship that would take them on their great adventure across the Atlantic. Little was their concern that the trip might be dangerous and lacking in creature comforts. Once the ship was located they began busily exploring every nook and cranny of what would be their home for the next month.

As their ship sailed out of the harbor, Martin led Mary to the stern. He gazed sadly at Wexford Bridge, silently remembering the scaffold where, after the Rebellion, so many of his fellow men lost their lives. Filled with emotion they lingered, arms entwined, until their beloved Ireland faded from sight.

CHAPTER TWO

Frampton—1820

On their arrival in Quebec, Martin arranged for a wagon to take them to the new development called Frampton, about thirty miles from Quebec City, to select the property for his farm. With the bogs of Ireland on his mind he chose the high land, as did most of the other Irish emigrants arriving about the same time. Unfortunately, the soil turned out to be hard and stony. The rich soil of twenty years earlier had eroded and washed away, leaving a rocky, unproductive soil. The French settlers wisely took the low land which was rich and loamy. Martin nevertheless planted crops of wheat and corn and by adding odd jobs in Quebec City in the winter was able to provide for his growing family. By this time the couple had been blessed by the births of John, Ellen (always called Helen) and Daniel.

Several families from Ireland joined the Murphys, including Patrick Martin with his sons, Patrick and Dennis (the latter became well known in California) and his daughter Ann, who became the wife of James Murphy. William Miller, a native of Frampton, married the elder Martin's daughter, Mary. About the same time, three members of the Sullivan family, John, Michael and Mary arrived.

The elder Martin was mild of manner yet with a presence that commanded respect. It was to him his companions turned for guidance. In essence he was their Irish chieftain and the Murphy home their gathering place. A frequent topic for discussion was their desire for a parish church. In 1825, he and the other Irish settlers signed a petition to the bishop for a church and in 1831 donated to a fund to bring a priest to Frampton. The settlers' prayers and efforts were rewarded by the building of St. Edoarde de Frampton Catholic Church.

Back in Ireland in 1828 the younger Martin turned twenty-one, enabling him legally to dispose of the farm. He and Margaret set sail from Wexford harbor for Quebec on board the *Thomas Farrell*, a packet ship. Under sail only three days, they encountered rough seas that caused damage to the ship, forcing it to make port at Waterford for repairs. Two weeks later the *Farrell* was again ready to sail, but missing were many of the wary passengers who, in the meantime, had had a change of heart. Not so with the spirited young Murphys, who eagerly boarded ship and enjoyed smooth sailing, crossing the Atlantic in twenty-eight days.

On their arrival in Quebec, they located Patrick Bolger who, meanwhile, had decided against farming in Frampton and settled with his family in Quebec City. That same year Margaret married a young Englishman named Thomas Kell. The younger Martin decided to remain in Quebec City motivated, no doubt, by having renewed ties with the Bolger family whose daughter Mary's company Martin enjoyed. The attraction between the two young people grew and they were married in one of the first non-French weddings in the French Cathedral of Quebec in 1831. Tragedy struck their happy marriage twice in Quebec when the couple lost two infant daughters, both under the age of two years, during a cholera epidemic.

In the hope of whole-heartedly pursuing farming as their livelihood the couple purchased two hundred acres near Frampton not far from his family's farm. Both were young, strong and energetic and with Mary's help Martin leveled the trees, hewed the timbers to build a house and prepared the land for planting.

Politically, in 1840, the Canadians were seeking self-government, but the wealthy pro-British group in upper Canada gained control over the French Catholic group in lower Canada with whom the Murphys were compatible. Inevitably, the English settlers received favored treatment. Besides, the patriarch had long been discouraged with conditions in Canada. The region lacked sufficient roads, making the transportation of his grain difficult, and the St. Lawrence River was frozen five months of the year. With the combination of a decline in wheat prices and the long cold winters, it seemed that it was time for a change.

About this time Martin, Jr. learned that the United States government had bought from the Indians a large parcel of land along the Missouri River in Holt County called the Platte Purchase. It bordered the Indian Territory on the far western frontier of the country. An agricultural paradise, it was located a few miles north of St. Joseph which was a growing community. Not only was the land cheap, but more important to this Irishman, it was a democratic country.

Once again the Murphys would be uprooted.

In this era, much of the traveling through the wilderness took place on the waterways rather than by horseback or wagon. Mary Foley immediately started planning and packing for a long journey by water over rivers, lakes and canals. Once under way, however, it proved to be more like a vacation. For the Murphy children, not only was it a lesson in geography but an exciting experience.

After boarding at Quebec they steamed up the St. Lawrence River to Montreal, crossed Lake St. Louis and crossed back again on the St. Lawrence to Kingston. They then crossed Lake Ontario and entered the Welland Canal that connected with Lake Erie at a higher level. Moving up the locks was a source of fascination for the younger Murphys. After crossing Lake Erie to Cleveland, Ohio, they traveled over another canal, the Fulton (which is no longer in existence) to Portsmouth, Ohio. They were now ready to start their long voyage down the winding, twisting Ohio River. At Cincinnati, a growing city, they took on passengers and freight and after another stop at Louisville, Kentucky they reached Cairo, at the Ohio's confluence with the Mississippi River.

From here they followed the Father of Waters, as the Mississippi was known, to St. Louis, a bustling community whose levee was lined with steamers. The younger Murphys were itching to see this thriving city that sat on high ground above the river, but there was no time—they had to transfer their belongings immediately to a wood-burning sidewheeler, one of many such steamers that ran between St. Louis and Kanesville (Council Bluffs).

Exhilarated to be heading west on the Big Muddy, as the turbulent Missouri was familiarly known, they were on the last leg of their journey—a five hundred-mile stretch to St. Joseph. Still, travel was slow going against a strong current—the shallow-draft boats loaded so deep with their cargoes of wagons, mules, horses, saddles, and cord wood for the engines, that the water came over the gunwales.

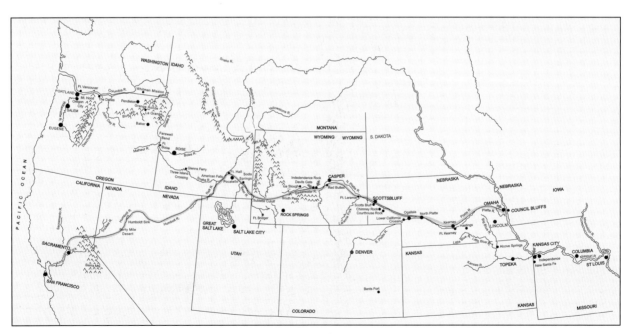

The route of the Stephens-Murphy Party, Council Bluffs, Iowa to California. FROM *TRAIL OF THE FIRST WAGONS OVER THE SIERRA NEVADA* BY CHARLES K. GRAYDON.

In contrast to the sameness of the Mississippi, the scenery on the Missouri was varied. Around a bend they might see a creek inhabited by a covey of ducks or, around another, a little cove with wildlife such as turkeys or geese. But mostly the shore was heavily wooded with an occasional open meadow or tiny village. The only man-made monument in view was the impressive white limestone state capitol standing majestically on a hill at Jefferson City. On its walls hung a portrait of Missouri's expansionist Senator Thomas Hart Benton whose son-in-law, John Charles Fremont, the immigrants would meet in California. The young Lieutenant Fremont followed the Murphys on the Missouri a year or so later on a thinly veiled exploratory expedition with overtones of the part he was to play in the Mexican War.

Eventually even the Missouri vistas became tiresome, but not so the capricious, ever-changing course of the river with its shifting sandbars, its banks wearing away on one side and building up on the other, and broken tree branches that conspired to form barriers presenting possible disaster at every bend. One of the biggest problems was the muddy water that had to be drawn from the river to make steam. It was so full of silt that it clogged the boilers.

More than once a heavy wind blew their boat ashore into a sand bank and each time the energy-charged Murphy boys, happy to have the freedom of touching land, pitched in with the crew to help dislodge it. There was nothing elegant about the Missouri steamboats—they were smaller, tougher, plainer—so flat-bottomed that those on the upper river drew only 30 inches of water.

On board, making up a heterogeneous passenger list, were French fur traders, card sharks who profited from such gullible subjects as the soldiers headed for Fort Leavenworth, and mountain men in their buckskins. In contrast to these travelers were the settlers from Kentucky, the Carolinas, Tennessee and Virginia, farmers all, who, like the Murphys, had been captivated by stories of the Missouri agricultural paradise with its fertile soil and cheap land.

Many immigrants had chosen Missouri over the Northwest because it was a slave state and they could bring their slaves. The young Murphys, who had never seen a black person before, enjoyed listening to their singing of spirituals. At the same time they could not fathom the thought that human beings could be bought and sold. Most of the Southerners disembarked at Wayne City, a muddy little river settlement. From

this point it was a short trip to St. Joseph where the Murphys believed their future lay.

The watery journey had become tedious but the family's excitement grew as they neared their destination. They arrived to find a few scattered farms and only a crude mill for grinding corn left by previous settlers who had named the place English Grove for the only cluster of trees that could be seen. As fellow countrymen joined the Murphys, they changed the settlement's name to Irish Grove.

Two years later the younger Martin and Mary Bolger followed with their sons, James, Martin, Patrick and little Barney, and were followed a short time later by the patriarch's second son, James, and his wife, Ann, with their infant daughter, Mary Frances. Then came the patriarch's daughter, Mary, with her husband, William Miller, and their three children, and John Sullivan with his younger brothers, Robert and Michael and his sister, Mary.

The younger Martin purchased three hundred ten acres next to his father and planted corn and wheat. James, on the other hand, eschewed the soil to work in the lumber business in St. Joseph. While the Murphys found the freedom that was so important to them, there was still a lack of schools and churches. True, the watercourses of the river area created a rich soil that produced bounteous crops. The newcomers were, however, alarmed to learn that the area was subject to the malaria virus.

The patriarch's beloved Mary Foley was the first in the family to succumb to the disease. As one of the bereaved Murphys said, "None knew her but to love her; none named her but to praise." In addition to losing the Murphy family matriarch, the younger Martin and Mary Bolger suffered the heartbreak of losing their nine-month old daughter, Ann Elizabeth, the following year.

Although those living at Irish Grove had heard glowing stories of the Rockies and the Pacific Coast from an old trapper named Robidoux who had a trading post at nearby St. Joseph, it was a visiting Jesuit priest working with the Kickapoo Indians in Kansas whose words made the difference for the Murphy family. When he learned about the many people in Holt County who were ill with malaria he came to nurse them and to console the bereft.

He told the elder Martin about California, an arcadia where there was perpetual sunshine, fertile valleys unfurrowed by the plow, and where grain grew wild. He explained that California was a Catholic country where the Franciscan padres built missions on "all the hills and in all the vales." That was enough to light a fire in the fifty-nine-year-old Martin Murphy who was despondent over the death of his wife and granddaughter.

He gathered his family around him and told them Father Hoecken's story. When he asked what they thought of going to California they were unanimously in accord. For his three unmarried sons, Bernard, John and Daniel, and his unmarried daughter, Helen, it offered a new adventure.

This was just the challenge the younger Martin and Mary Bolger needed. Suffering over the loss of Martin's mother and their infant daughter, they were concerned about their other four children. Also electing to go along were James and Ann Murphy and her Martin relatives: Mary Murphy and her husband, William Miller, with their three children, and John Sullivan, with his brothers and sister.

Losing no time, they started to liquidate their assets—selling their property and investing in wagons, oxen, horses, farming implements, tin ware, cast-iron skillets, staples, water kegs and what little furniture they could squeeze in for a trip of eight or nine months, not knowing what they would find in far off California. They had many meetings to attend and decisions to make such as choosing their departure point—whether it should be Independence or nearby St. Joseph or Kanesville. A vote was taken and Kanesville it was. The elder Martin, with a passport secured from Missouri Governor Thomas Reynolds, headed the group of 27 Murphys, including in-laws, children, and the Sullivans.

CHAPTER THREE

Kanesville (Council Bluffs)—1844

Bursting with anticipation, the Murphy party arrived at Nishnabotna, fifty miles below Kanesville (later to become known as Council Bluffs) at the western border of the Iowa Territory. Their trip from Irish Grove had been plagued with rain, rain and more rain that wouldn't go away, forcing them to ford swollen streams in eastern Kansas and to make their way through mud axle-high. It was, in fact, Missouri's wettest spring on record.

At this point the Murphys were facing the unknown; wild rivers to cross, trackless deserts in the heat of summer, and native peoples who might confront them. But their greatest challenge would be conquering the mighty Sierra Nevada and taking their wagons and cattle into California, a feat never before accomplished.

The elder and younger Martins had put a lot of thought into their plans, timing their departure so there would be grass on the prairie to feed the cattle and yet time to get over the mountains before the snow fell. Their wagons (the larger ones called prairie schooners) were sturdy, their oxen and horses healthy and the food supplies ample.

Father and son saw to the medical supplies, as well as to the utilitarian needs of farming such as plows, axes and shovels. Rope was needed to pull the wagons across rivers, for keeping the wagons level on inclines, or pulling them out of the mire. Chains would be needed for locking the wheels on steep downgrades.

In the event they might need to defend themselves from Indians, they brought ten pounds of powder, thirty pounds of lead for making bullets and two hundred percussion caps.

Added to all that was the bucket of tar, hanging from the rear axle (later substituted with animal-fat) for greasing the wheels and any other parts that might be needed.

Mary Bolger, always practical, had organized the food with dried staples, beans, flour and cornmeal, along with pilot bread, bacon, coffee, tea, sugar, salt and dried apples (a preventive against scurvy) and even a milk cow for the children's milk. To take advantage of the surplus milk and the movement of the wagon, a butter churn was attached for making butter. She planned not only for her immediate family, but also for her father-in-law and his four unmarried children as well. His youngest daughter, Helen, would help her with the cooking and the combined families would enjoy their meals together.

Camping with their wagons close by as they had done on the journey from St. Joseph were family members Jim and Ann Martin Murphy with their three-year-old daughter Mary Frances, Ann's father, Patrick Martin, and her two brothers, Dennis and Patrick. Also camping nearby were Mary Murphy Miller with her husband, James, and their three children. After the evening meal was eaten and the children were put to sleep the Murphy clan would gather around the bonfire to sing, dance and exchange yarns about the day's events.

On the banks of the river at Nishnabotna they found a large assemblage of pioneers busily preparing their wagons for the journey west. The elder Martin visited among them and, after listening to their plans and the decision on the route that they would be taking, called for a family council. He proposed that the Murphy family join a larger train, there being safety in numbers, especially in case of an encounter with unfriendly native people. The "aye" vote was unanimous. The extended Murphy family signed on with a group going to Oregon and these two groups were joined by another going to California. Eleven wagons altogether were going to California. Three of the groups were planning to take the Oregon Trail and it was agreed that the two trains would part company at Fort Hall.

At dawn on May 18, 1844, after a prodigious coordinating effort, the combined California and Oregon parties, totaling twenty-eight wagons, started up the Missouri River single file. Each day scouts were sent ahead, and it was decided that the person in the lead would, the next day, take his place at the end of the line and that the younger men would "herd-up" the loose cattle in the rear.

Before reaching Kanesville the parties camped on land belonging to the Patowatomi tribe. The chiefs and the great men of several tribes, the Otoes, the Sacs, the Foxes and the Pawnees, were camped there to smoke a pipe of peace with Pa-reesh, the Patowatomi chief.

The Murphys, especially the children, were fascinated with the native customs and dress—the war paint, bear claws and beads around their necks, their earrings and nose ornaments. Most had feathers inserted in the topknot of hair on their otherwise shaved heads while others had their long hair braided and decorated with brass plates. Yet the Americans' amused expressions turned wary when they witnessed a war dance and saw the braves raise their tomahawks, lances, bows and arrows, spears and war clubs—each time letting out weird-sounding yells—all this accompanied by the beating of a drum and the singing of war songs.

One of the Oregon members, A.C.L. Shaw, presented two sheep to the chief to pay for his ferrying fee. When Pa-reesh told a Pawnee chief about it the latter made a speech (translated by the guide, Caleb "Old" Greenwood) expressing his wishes that they would travel safely through his nation. From time to time, when the chief struck his breast, his warriors would respond by raising their voices.

The next day the joint wagon trains met to establish rules and regulations. This was accomplished but not without considerable wrangling which, unfortunately, would prove indicative of the temper of times ahead. On the agenda was the election of a captain. A likely candidate, because he was the most educated, might have been Dr. John Townsend, a native of Fayette County, Pennsylvania. Townsend joined the expedition as a licensed physician and part-time farmer with nomadic tendencies. He had moved first to Ohio, then Indiana and finally had come from Missouri and joined the group. In his party were his wife, Elizabeth, her younger orphaned brother, seventeen-year-old Moses Schallenberger, described as "callow youth, not yet dry behind the ears," whom Townsend had adopted, and a fourth member, his hired hand, Francis Deland.

Dr. John Townsend was a leading member of Murphy party. COURTESY HISTORY SAN JOSE.

Another possible candidate was sixty-four-year-old Isaac Hitchcock, an experienced trapper with a knowledge of the West, whose widowed daughter, Lydia Patterson, and her five children made up his party. The Murphys with twenty-four family members (including in-laws, plus their extended family) and the four Sullivans, who had been with them in Canada and at Irish Grove, added up to twenty-eight—over half the total party. They could have controlled the voting in their favor but declined. Neither the elder Martin, nor the younger Martin, who, by this time, was sharing family leadership with his father, had any experience in the West. The Murphys also realized that, because of their overwhelming numbers, they might be accused of partisanship and thus cause dissension. Still, with their innate ability, they would become leaders.

They cast their votes for a gritty mountain man and former Indian agent, Elisha Stephens, (sometimes misspelled by historians as Stevens, although recorded documents indicate he signed his name Stephens). A forty-year-old native of South Carolina, raised in Georgia of Huguenot stock, he had hunted and trapped in the Northwest for the fur trade. During the time the emigrant train was preparing for their big move, he was working as a blacksmith with the Indian subagency at Kanesville and became caught up in the California dream. A homely man with narrow eyes placed too close together, a hawk nose and large ears, he was a loner lacking in communication skills. In retrospect, he seemed an unlikely candidate for the job of wagon master. The Oregon contingent, the larger of the two companies, had good men such as John Thorp they could have proposed, but they went along with the selection of Stephens—a decision some of them later regretted.

Elisha Stephens was elected captain of the combined parties going to Oregon and to California. COURTESY STOCKLMEIR LIBRARY AND ARCHIVES, CALIFORNIA HISTORY CENTER.

A small group of the Oregon contingent had rejected the election of "Old Greenwood," to be their guide. They considered him too old and almost blind. Also they may have been turned off of Greenwood because of his long hair and buckskin clothes. Called the White Chief of the Crows, he had become more Indian-like than white man after living among that tribe for thirty years with his wife, Batchicka, daughter of a French trapper and a Crow Indian woman.

Some of the Oregon-bound members not only refused to vote for Greenwood, but could see no need for a guide and refused to pay their share for his services. Obviously, they had not given enough thought to the fact that they would be crossing a two thousand-mile wilderness that was unknown to them and that Greenwood had trapped all over the Great Basin. Even more importantly, he had knowledge of the habits and language of the Indian tribes they were sure to meet along the way.

This was the comeback trail for the fur trapper and mountain man, eighty-one-year-old Greenwood, who said he had been as far as Fort Hall and claimed to have been in California. Several years earlier he had become blind. When the Crow medicine man was unable to help her husband, a determined Batchicka built a canoe and took him with their six children, including a brand new baby, down the Yellowstone and Mississippi rivers to St. Louis where a "white man" doctor performed surgery.

Miracle of miracles, he could see. Apparently, the good doctor had removed cataracts. But there was a price. His loyal Batchicka became ill from over-taxing her strength so soon after giving birth and lived only a short time longer. A few years later, with his knowledge of the Great Basin, Greenwood decided to become a guide. With him were his two sons, eighteen-year-old John, and seventeen-year-old Britain.

Making up the rest of the California party were Allen Montgomery, a gunsmith, and his wife, Sarah, from the same Missouri town as the Townsends; Joseph Foster, a vigorous young man in his middle twenties; four single men working their way as drivers of oxen or cattle herders, Matt Harbin, a Tennessee native filled with the wonders of the distant land, California, and a hankering to go west; Olivier Magnent, John Flombeau, a Canadian of mixed French and Indian blood, Vincent Calvin and Edmund Bray, an Irishman.

Two days later they met their first big challenge, the crossing of the Missouri River. Chief Pa-reesh sent his flat-bottomed boat to help ferry their wagons, but that was the easy part. The night before a nasty midwestern storm had descended upon them. The wind blew, the rain came down in sheets, thunder roared and lightning flashed.

Captain Stephens, probably concerned about time limitation to cross the Sierra Nevada before the snows, the next day ordered the company to swim the cattle. Some of the Oregon-bound wagon owners refused. The poor beasts, terrified of the swirling, white-capped water, refused to budge. When pushed into the river they tried to save themselves as they bawled in fright, going around in circles and jumping on each other's backs. Many became mired in the sand that had become firmly packed. It looked for a while as if they might have to abandon all their cattle. Chief Pa-reesh volunteered to help by pulling with ropes and digging with shovels.

At the younger Martin's suggestion, they waited until the river receded. Then he and his brother James, working from a canoe, tied a rope around the horns of a more docile ox and by coaxing and pulling succeeded in getting him far enough into the water so that he had to swim. With a little more coaxing the others followed.

A correspondent from the *Daily Missouri Republican*, in the June 14, 1844 edition, considerably understated the case when he reported (from Oregon, Holt County, Mo., May 30):

"A small party of emigrants crossed the Missouri at Council Bluffs last week after electing Mr. Stephens, formerly a resident of the Indian country, Captain. They crossed the river without difficulty or loss, except for a few loose cattle. There were twenty-seven wagons in all, about forty men and a large proportion of women and children."

They paid their ferry fee to Chief Pa-reesh and before they continued on their way the chief gave each immigrant a bag of salt. They were now ready for the next obstacle on the course. A day after Bellevue they started up Papillon Creek—stopping in the morning to build a bridge across the creek. The rain returned that night and continued the next day and the day after. Tempers became tight.

One of the most vocal of the Oregonians, A.C.L. Shaw, took out his frustrations on Captain Stephens. Still grousing over the Missouri River crossing, he accused Stephens of giving orders that did not conform to wagon train regulations. In an excited state he rushed into camp, picked up his gun and fired, reloaded and declared he was going to kill the captain. Joined by others who shared his feelings, an election was called. Those voting for Stephens the first time repeated, and he was re-elected. They continued their journey and, while traveling the Loup, tempers were short. Stephens resigned his captaincy in a huff—only to change his mind a short time later, claiming he was put up to it.

Always of concern was the possibility of native people raiding their livestock. The cattle were under the charge of a herdsman until dark, after which the animals were chained to the wagons. They had fewer qualms about the cattle when they came to a Pawnee Indian village and found only women, children and elderly men. It turned out that the able-bodied men had been victims of a Sioux attack. According to Greenwood, the Sioux were considered the most warlike, cruel and treacherous of the tribes and were the most feared by the pioneers.

When they reached Otoe Indian country, Old Greenwood advised Stephens that this tribe, while friendly, had a reputation for not always being honest. As a safeguard he suggested that at night they draw the wagons into a circle and place the tongue of one wagon on the hind wheel of the one in front to create a fortification. A guard was stationed outside with a replacement every two hours. John Murphy and Moses Schallenberger, still in their teens and by this time bosom pals, were chosen corporals of the guard to go from sentinel to sentinel to check to see if each man was alert. The cattle were under the charge of a herdsman until dark after which they were brought into the corral and chained to the wagons, a cautionary move to protect against prowling wolves.

Time passed slowly and about midnight a restless John suggested to Moses that they play a trick on John Sullivan—a break in the monotony of the daily routine. John released some of Sullivan's cattle, and after driving them into the woods gave an alarm. Sullivan, who had been apprehensive all day convinced that Indians were hovering about the camp, grabbed his gun and all joined him in pursuit of his oxen.

After a chase, the alarm went off again—the cattle went farther away this time making it necessary to follow them by the clinking sound of the yoke ring. At one point Sullivan climbed atop a log to listen for the sound. John, lying concealed behind the log, fired his shotgun into the air.

By this time, Sullivan's imagination was running wild and he hurried full speed back to the wagons crying out that he had been shot at by an Indian. Meanwhile, John and Moses had recaptured John's cattle and tied them to Sullivan's wagon. Taking no chances,

Sullivan stood guard over them until daylight. The next morning the captain complimented the boys for capturing Sullivan's cattle, at the same time wondering why the Indians had taken Sullivan's each time. John and Moses glibly replied that his cows were white and more easily seen at night.

A few days later the wagon train reached a tributary of the Elkhorn River. This time the cattle swam across without any problem, but the water level was too high for the wagons. Not even a primitive flat-bottomed boat was available to serve as a ferry. It wasn't easy but the travelers solved their problem by emptying and dismantling all the wagons and loading them, one at a time, on a pair of improvised ferryboats they had made by sewing rawhides together and stretching them tight over the wagon beds. They were then pulled across with rawhide ropes and tied around trees on each side. This innovative process of making rawhide covering proved to be helpful to them later on. There would be eighteen more rivers to cross and re-cross.

The women and children were next, but those in charge were in trouble when they tried to move a whole wagon. It was top-heavy and the boat and the wagon floated down the stream some distance before it could be reached and pulled ashore. Their next project was to improvise bridges across two sloughs. While crossing one of the bridges the younger Murphys' wagon turned over resulting in all their possessions getting drenched. Fortunately it was a sunny day and the company suspended travel for the afternoon so the Murphys could put their things out to dry. Martin and Mary Bolger were embarrassed and apologetic for having caused the delay. Still, they could be thankful that no one was hurt.

As the caravan continued across the rolling, barren plains of Nebraska, the children amused themselves watching the prairie dogs sitting at the mouths of their burrows, holding their paws in front of them as though in prayer. Soon the Murphy elders spotted American pronghorn in large numbers. John and Dan Murphy, both superb horsemen as well as excellent shots, met the challenge of this fleet-footed animal and provided the Murphys with meat for many a meal.

From the Elkhorn the emigrants reached the muddy Platte River, which was facetiously said to be "too thick for soup and too thin to slice." Following the north side, they were on their way across Nebraska to Fort Laramie—their route later to be called the Mormon Trail for a group who followed. Along the way, a division formed within the company. John Thorp was

elected to lead the Oregon-bound group until they reached the Willamette River. Meanwhile, one of the Oregon group took ill and became the only fatality of the long journey. A critical member of the Oregon group attributed the death to Dr. Townsend insisting that it was the medication he gave the man.

Adding a little excitement for the emigrants on their, so far, monotonous crossing was their first sight of buffalo, a dozen or so running abreast, their shaggy heads held high. The ebullient threesome of Moses, John and Dan, who couldn't wait to get their first buffalo, started out after them. Unfortunately, they expended an unworthy amount of ammunition to bring down a scrawny old bull that had been driven from the herd by the younger bulls. The tough, strong meat was a disappointment but when, subsequently, younger animals were downed, all agreed their meat was the best anyone had ever tasted.

Soon after, they experienced an awesome sight that would leave lasting impressions on the young Murphy minds. A vast herd of thousands of buffalo on a wild stampede suddenly appeared on the horizon headed in their direction. The emigrants quickly goaded their oxen into a run, averting the onslaught by a matter of yards. The herd was more than an hour in passing, leaving a cloud of dust as they plunged into the Platte River.

Along the way the ground was covered with bleached buffalo bones much to the dismay of the Indians who valued the buffalo meat as their principal means of sustenance. The skins served as material for their clothing and a covering for their wigwams. Small wonder they feared the breed would be killed off. For their part, the natives were careful never to kill an animal as long as they had any meat.

This was one of the Indians major grievances against the Americans who slaughtered the buffalo by the thousands for their skins. On one occasion during the Bidwell-Bartleson party's crossing, a member brought back the tongues of eleven buffalo he had killed, boasting he had left the corpses on the prairie for the vultures.

As they continued along the treeless prairie of the North Platte, the summer heat was suffocating. The Murphy women were thankful for their sunbonnets and the men for their broad-brimmed hats. While not pushing hard, the train was making excellent time. They were to witness a phenomenon of this desert—a hailstorm such as they had never before experienced. Balls of ice, ranging from the size of a walnut to an egg, filled the atmosphere to the accompaniment of ominous thunderclashes and flashes of lightning shooting across the sky. Had one been hit on the head by a hailstone, the results might well have been fatal. Members of the train ran for shelter and the frightened teams reared until the drivers seized control. When it was over some fifteen minutes later, the emigrants were grateful no one suffered severe injury.

After that frightening interlude, the wagon train returned to the monotony of prairie and more prairie. The parents, in trying to find diversions for the children, told them about a phenomenon of nature they would soon be seeing called Chimney Rock. The younger Martin's boys, James, Martin and Patrick, made wagers as to who would see it first. But it was their three-year-old brother, Barney, who first spotted the imposing figure of a spire, composed of tightly compacted sand and clay, rising above a cone-shaped base two hundred feet above the plain. It was best described as resembling an inverted funnel.

Plagued by swarming gnats and the poisonous bites of mosquitoes, the wagon train members—especially the elder Martin, who was allergic to them, were naturally eager to hurry on. As the country became hillier, the streams more rapid and the air clear and dry, the emigrants knew that Fort Laramie, the gateway to the Rockies, could not be far away. They were passing Scotts Bluff, another dramatic sight with its rock formations rising four hundred feet above the ground like a mythical city.

They were now in Wyoming. When at last they approached Fort Laramie, they beheld an imposing structure in the shape of a quadrangle, its walls six feet thick and fifteen feet tall with pickets between large watchtowers at the corners. The fort was situated on a rise twenty-five feet above the river with the Black Hills and snow-capped Laramie Peak rising in the background.

At Laramie the company had its first contact with a white settlement since leaving Kanesville eight hundred miles ago. Within these whitewashed *adobe* walls resided fifteen or so men, mostly French trappers, their strikingly beautiful Sioux wives with pale, copper-colored skin, and their numerous mixed-blood children. At this time it was a fur trapper's trading post, built by the American Fur Trading Company. Five years later it was sold to the United States government for a fort.

With ample feed for their animals, the party decided to spend several days resting, stocking up on buffalo meat and having some blacksmithing done. The men were uneasy, however, when they saw about four thousand Sioux camped in their wigwams around the fort. They remembered Old Greenwood's admonition that the Sioux were less than hospitable to passing wagon trains. Greenwood assured the Murphys that, inasmuch as the Sioux party included wives and children, they would not threaten the emigrants.

With his usual ease in interacting with the natives, Old Greenwood went down to visit their camps and to make friends with the chiefs. He was thus able to pave the way for the emigrants, whose shoes were wearing thin from the long walk, to barter with the natives for moccasins. They also traded horses for Indian ponies, believing they would be more adaptable to the rough terrain ahead. Their dealings with the Sioux came off without incident, but the Murphy-Stephens party members were apprehensive as they moved on, fearing that after they left their families the Sioux might engage in a war party.

The wagon train had not gone far before they had another meeting with a large tribe who appeared to be in a war-like mood. Old Greenwood again worked his diplomatic skill. After the emigrants drew their wagons into a defensive circle, brought the stock inside and had their guns ready to attack, Old Greenwood rushed in, cautioning them sternly to hold their fire. At the same time he held back his sons Britt and John, knowing full well their hostile feelings toward other tribes. Raised with the Crows, they were taught to believe the Crow people were the aristocrats of the Indian world. Greenwood approached the natives using sign language to indicate he wanted to be friendly. When they saw that he understood their language and customs they negotiated a settlement. After smoking a peace pipe, one of many he smoked on the long trip, the emigrants were permitted to proceed unmolested. Had it not been for Greenwood they might have lost their stock and possibly even their lives.

The warriors followed the wagon train for several days. One of the young braves, having taken a fancy to the elder Martin's daughter, Helen, told Greenwood he wanted to barter for her. Greenwood told him she was not for sale. The warrior was insistent, but when he became convinced of her unavailability, he accepted gracefully. In fact, he gave her several presents. The elder Martin had been more than a little apprehensive while all the negotiating was going on for his youngest and favorite daughter. Still, it ended well with no hard feelings on either side. Helen thought it quite amusing and thanked the young brave with her most beguiling smile. Her pretty face and sparkling personality would later charm prominent visitors to her father's *hacienda* in California, many of whom wrote about her in their memoirs.

The scenery changed when they started climbing the winding Black Hills. In contrast to the bleached plain they had just passed through, the hills were covered with a heavy growth of juniper and pine. Mary Bolger and Helen were happy when they found wild pear and peas to vary the menu.

On the Fourth of July the future Californians arrived at Independence Rock which had received its name from a group of mountain men who had celebrated Independence Day there fifteen years earlier. The stop was well-timed for Mary Murphy Miller, because it was on this day that she gave birth to a baby girl. Dr. Townsend performed the delivery, assisted by Mary Bolger who, having produced seven babies, knew first-hand the pain of labor and the joy of birth. In honor of the date and place they named the baby girl Ellen Independence Miller.

The patriarch, proud of his American citizenship and the meaningful date that marked the emancipation of the American colonies from England, compared their situation with Ireland's—the only difference, he said, was that the Yankees had succeeded where Ireland had failed. He called for a celebration of Ellen Independence's birth and of the Fourth of July with a roast of their finest cuts of buffalo meat.

The Rock itself resembled an enormous granite whale or monster and, according to the measurements taken by Lt. John Charles Fremont two years earlier, was six hundred yards long and rose one hundred feet above the ground like an island. The Murphy boys and the Miller children added their names to the hundreds already scratched on it by fur trappers, travelers and pioneers who had tarried along the way. Even Father Peter de Smet, the famous Jesuit missionary of the Northwest, who best described the rock as the "great registry of the desert," etched his name on it as did William Sublette, who started it all when he "discovered" the Rock in 1830.

Some of the names inscribed on the Rock the Murphys would get to know well later, in California. They included Thomas Fitzpatrick, one of the foremost guides of the West, credited with the discovery of the South Pass and always known as "Broken Hand" because of an injury caused by a gun going off in his hand. Another was Charles Weber, founder of Stockton, who will one day marry Helen Murphy. The names of John Bidwell, who becomes an important figure at Sutter's Fort; and Father Gregory Mengarini, S.J., who, only a few years hence, will teach the Murphy children at Santa Clara College, which their father helps found, were also there. Then, of course, there was John Fremont, who carved a cross, considered the symbol of the Christian faith. The symbolism of this carving haunted him as anti-Catholics opposed him in his campaign for president, even though he himself was an Episcopalian. He and his scout, Kit Carson, only two years later, will camp at the younger Martin's *rancho* near Sutter's Fort.

Beyond the Rock, the emigrants struck the winding, icy cold Sweetwater, with water as clear as the Platte was muddy, and, for a welcome change, it was a scenic trip. From time to time they passed green valleys and a profusion of flowers so pleasing to nature-loving Mary Bolger. Traveling up the stream they passed another famous landmark—Devil's Gate—at which point the river, as though in a mad rush, came charging through a cleft in the mountains that rose up on either side. They found ample elk and other game and while crossing and re-crossing the stream frequently stopped to catch carp for dinner. When the snow-capped Wind River Mountains came in view, the icy wind from the mountains and a driving rain signaled a change in the weather. After leaving the Sweetwater, the grade was so gradual and the travel over the wide road so easy they scarcely realized they were in the Rockies. When they came to a pass where the mountains rose fifty to sixty feet on either side at an elevation of seven thousand five hundred feet, they recognized John Fremont's description in the report on his expedition to the Rockies in 1842. They had reached the South Pass of the Rockies—the great Continental Divide. This was an important landmark on their nine hundred-mile journey from the frontier. It was the halfway mark across the continent—sometimes referred to as the "backbone of Uncle Sam."

When they reached Pacific Springs and saw the water running toward the west, some of the party, with little knowledge of geography, was elated, thinking the journey was almost over. They little knew the worst was yet to come.

The emigrants headed west across the sage-covered plateau until they came to the Little Sandy River and, a dozen miles farther, the Big Sandy, where they spent the night. The big question facing them, and one that caused considerable discussion, was whether they should play safe by staying with the established route south by way of Bridger's Fort, or whether they should try a cutoff suggested by Isaac Hitchcock—one he had heard about when he was in the Great Basin ten years earlier as a fur trapper. Hitchcock warned them the route had never before been followed by wagon trains and that there would be no water. Still, he thought they could get through without danger and in this way save several days.

Aside from the timesaving aspect, this route led to a crossing of the Green River (later known as Mormon's Ferry) down a steep canyon between high, sheer bluffs. It was the only point where wagons could get down to the river. After much discussion between the Murphys and Stephens, they decided to rely upon Hitchcock's knowledge of the country and sense of direction. Without a trail or landmark, under an oppressive August sun, they started across the sixty-mile stretch of arid, sandy desert through rugged ravines and dried-up alkali lakes.

Unfortunately, Hitchcock had not considered the slow pace of the oxen. The emigrants, for the first time, misjudged the amount of water to carry and brought insufficient cut grass for the oxen. Traveling from dawn to dusk over dry, dusty, rough country, they were just as weary and thirsty as the cattle. Nerves were on edge, but at Hitchcock's insistence they continued through the night.

By late afternoon the next day a breeze sprang up and the cattle got a whiff of water. That was all they needed. Half-crazed with thirst they started to run, dragging the wagons. The wagon leaders, in panic, had to release them in order to prevent serious accidents and the destruction of their wagons. The animals ran to the river where they were later found belly-deep in the water among the willow trees. The cattle had suffered terribly, and during the night about forty head wandered away.

The shortened route was a miserable experience but it did save them five days. The shortcut became known as Greenwood's Cutoff and later as Sublette's Cutoff. On the west side of the Green River they ascended a narrow, dry gulch with a wall of sandstone rising up on either side. As later emigrants taking the Greenwood Cutoff inscribed their names and dates, it became known as "Names Hill."

The next morning, six young men were assigned to go in search of the missing cattle. As they started off they were of different minds as to where to find them, so they separated—one group headed back to the Big Sandy while the other three started for a designated point on the Green River some twelve miles south. The trio, headed by Dan Murphy who, although the youngest, always seemed to be the leader, were moving along single-file when Dan, riding in advance, suddenly spotted the feathered topknots of Indians. Dan ducked his head, threw his body to the side of his horse and signaled to the others to do the same.

After following him full speed for one-fourth of a mile to a small canyon, his companions asked Dan what the matter was. He whispered, "Indians." They secured their horses to nearby bushes and, lying on their stomachs, crawled to a point overlooking the plain where they saw below a war party of a hundred or so Sioux who were so near they could hear them talking. After the last feathered, top-knotted Sioux had left, the three heaved a sigh of relief. It was a call too close for comfort. Mounting their horses they continued on to the Big Sandy where they found the missing cattle and spent the night.

But their troubles were not over yet. The next morning, after rounding up the stock, they set out for Green River. About a mile away they spotted two mounted Indians on the top of a hill. Soon they saw two more and then more and more until they were surrounded by two hundred whooping it up and charging in an ominous manner that brought terror to their young hearts. They put their heads together and planned their strategy—each would shoot an Indian and would then charge through their camp. But just in case this didn't work, as Moses Schallenberger wrote in his account of the Murphy-Stephens party, they said a poignant goodbye to each other.

About twenty Indians advanced to within two hundred yards of the three nervous young men, who gave them a sign to halt. The braves sent three men without arms to parley. Within a distance of fifty yards, much to the emigrants' relief, the natives stopped and held up their hands in a sign of friendship. They were a party of friendly Snakes in pursuit of the same Sioux the wagon party men had seen the day before. Their new friends even assisted the three men in finding their cattle and driving them to Green River, where they rejoined their party.

The emigrant party now traveled up a high divide and, following a stream marked by willows, passed down to the beautiful valley of the Bear River. This turned out to be one of the most pleasant experiences of the trip. They found an abundance of grass for their stock, willow brush for their fires, trout for fishing and ducks and geese to shoot.

Finally they came to tall bluffs that were so close to the water they had to detour into the Bear River Mountains. Returning to the river they found old "Peg Leg" Smith, one of the first white trappers in the Rockies, famous for having amputated one of his legs with a butcher knife after gangrene had set in. He made a wooden leg for himself, buckled it on and it was, he said, "as good as ever." Living alone (he later took a Native American for a wife), an outgoing, hospitable man, he seemed to like company.

During the cold winters Smith, innately generous, often fed the Indians from his ample number of cattle.

The meeting with Smith proved to be profitable for the Murphys, because he not only gave them helpful advice, but exchanged some of his fat ponies for their poor, tired horses. As a result of this encounter, Smith, on occasion, visited Elisha Stephens at his Cupertino, California ranch at the site of what is now known as Blackberry Farm.

With fresh ponies, they proceeded down the Bear River and left it at Soda Springs to travel overland to Fort Hall.

CHAPTER FOUR

Fort Hall, August 10, 1844

The California contingent of the Murphy-Stephens party arrived at the Hudson's Bay Company Trading Post at Fort Hall on August 10, followed some days later by the Oregon-bound group who had come the longer route by way of Fort Bridger. Thus far the California-bound had taken a route that was more or less known to trappers. With Greenwood as pilot and Isaac Hitchcock on board the route had posed no serious problem. Although Greenwood's contract had terminated in the Rockies, he continued as guide and interpreter.

The party had hoped, when they reached Fort Hall, to obtain information about the journey ahead, but to no avail. All they could learn was that the Chiles-Walker party had passed through the year before and that at this point they had split—Joseph Chiles taking his group to California by way of Idaho and Oregon.

Joseph Walker headed south and, fortunately, the Murphy-Stephens party would be able to pick up his wagon tracks from time to time as far as Mary's Sink (later changed by Fremont to Humboldt.) What they didn't know, and probably just as well, was that Walker's party had later abandoned their wagons near what is now known as Owens Valley. With great hardship they crossed into California and, by a devious route across the upper Salinas Valley, eventually reached John Gilroy's rancho where some of the group stayed on.

On August 17, the Murphy-Stephens party, after replenishing their stores, repairing the wagons and resting the animals, struck out into the unknown, putting all their trust in the two older men, Greenwood and Hitchcock. They resumed their march southwest down the Snake River until they reached the Raft River (Bear Creek). At this point, they had their last farewell with the Oregon group who had joined them this far. In the course of their long journey the two groups had had their moments of disagreement, but close friendships were made, and it was an emotional parting—especially for the gregarious Murphys.

For two days the Murphy-Stephens party traveled along the Raft River. Then, for the next eight days, they proceeded westerly across a barren, sage-covered plain, passing the City of Rocks and Goose Creek to the headwaters of Mary's River (also renamed the Humboldt by Fremont). Here they found not only plentiful grass but also water high enough to dilute the alkali that would give later emigrants problems.

At this point they met, for the first time, the Shoshone Indians ("Digger" is a derisive name for them referring to a diet including roots). The most roughly housed of the tribes encountered on the trip, they lived in wickiups and brush shelters and fed on fish, grasshoppers, rats, roots and grass seed. Their language was practically unknown to Greenwood, but he was able to communicate with them by sign language and a few words of Shoshone. Although they appeared indolent, they were friendly—as it turned out, almost too friendly.

In large numbers (one diarist said hundreds) they visited the camp every night for the next ten days. At first the women and children of the wagon train were frightened, but with Greenwood's ability to communicate all went smoothly. The emigrants, nevertheless, continued their vigilance with guards on duty every night, and passed through the territory without problems except for a pony stolen from the elder Martin.

In the years that followed, the Shoshone did not fare as well with emigrants along the Mary's. Insensitive to the value of human life, the Americans amused themselves by shooting at the Shoshone from the back of their wagons as they would shoot at rabbits. The Shoshone learned to retaliate by shooting arrows into the emigrants' cattle, so that they had to be left behind for the Indians' own consumption. The situation grew even worse during the Gold Rush when the Shoshone retaliated by attacking small companies of emigrants.

Although the grass along the river provided ample feed for the animals and the heavy winter rains had reduced the alkaline in the river water, the oppressive September sun bore down unendingly. Each day of the five hundred-mile journey down the Mary's was a replay of the previous day. The party members no doubt would have agreed with Horace Greeley's assessment of it as "the meanest river in the world."

Up at dawn, after the cattle guards reported "all clear," the lead driver lashed a long bull whip without touching the newly yoked oxen. They knew the meaning of the sound and leaned against the yokes causing the wagon to lurch forward. When the call "catch up" went out, the other wagons followed single file. Before the sun was directly overhead they stopped for nooning, something to eat and a couple hours of rest.

By late afternoon, as they continued rolling along, the adults became hot and edgy from the constant creaking sound of the wagon wheels, the baying of the cattle bringing up the rear and the constant "gee" or "haw" call of the drivers. The children, now restless and cross, received spankings. Even the usually upbeat Mary Bolger couldn't bolster her family's spirits.

In the early evening, as the sun moved toward the west and began to shine in their eyes, the scouts would start looking for a spot to camp where there was water. As they had done so many times before that it became automatic, they unyoked the oxen and placed the wagons in the usual circle. The Murphy children scampered around to collect brush and whatever else was available for the fire while Mary Bolger and her sister-in-law, Helen, started preparing to cook supper. Tensions relaxed after the children were put to bed in the wagons and the older folks sat around the fire singing while the younger adults danced.

Finally, after having to digress on one occasion by way of a canyon which forced them to spend a night sleeping in the hills and on other occasions because of necessary detours, they reached Mary's Sink on the first of October. Their tired, bored expressions turned to smiles as they saw before them a wide, spreading meadow. Not only did that mean feed for the cattle, but ample water. The party, animals as well as people, were surprisingly healthy.

Cattle were let out to graze; the horses unsaddled. The elder Martin recommended and Stephens agreed that they spend a week there preparing for the challenge ahead: finding the way to and over the Sierra Nevada and on to Sutter's Fort. The men busied themselves with repairing outfits and mending damaged vehicles while the women cured meat and washed and mended clothes.

Excellent shots, the younger members Dan, John and Moses provided a diversity of menu with wild geese, ducks, sage hens, pronghorn and deer, all of which they found in abundance.

Moses Schallenberger. The night before leaving the Humboldt Sink, eighteen-year-old Moses discovered one of the Indians had stolen his halter, and he went for his gun. Martin Murphy, Sr. smoothed things with generous gifts to the tribe. COURTESY HISTORY SAN JOSE.

The Murphy-Stephens party had an even more serious problem. Having reached the Sink they now faced the moment of truth—where to go from here? There would be no more Walker wagon tracks, for he had turned south. The two Martin Murphys, James Miller, and Captain Stephens consulted anxiously over this question. Again they turned to Old Greenwood, who was the only one who could possibly interpret the verbal language and gestures of the Paiutes. He visited their camp and with his usual diplomacy and understanding, he quickly won their confidence.

When he asked to see their chief, they directed him to a middle-aged man who they said had been beyond the mountains. Greenwood didn't know the Paiute language but he and the chief were able to communicate with each other through sign language and drawing diagrams in the sand. The chief told Greenwood of a river fifty or sixty miles west of where they were camped that flowed from the mountains. Along this stream were large trees and good grass. If they followed this stream he said, they would come to a pass across the mountains.

Truckee, as the emigrants now called their new found friend because he used the Shoshone word meaning "all right" so frequently, volunteered to show them the way. With the Paiute chieftain as their guide, Captain Stephens, Dr. Townsend and Joseph Foster rode out with him for a trial run to look for the river and explore the country. On their return three days later they reported it was just as Truckee had described it.

A long time later the Murphys learned an explanation for Truckee's extraordinary friendliness. It seemed his tribe had preserved a legend that the world had begun with two couples—one black and the other white. After a while they started to quarrel. The great father was angry and told them they would have to separate until they could live together in peace. The white couple vanished—but, according to the myth, promised one day they would return. To Truckee the members of the Murphy-Stephens party were his long lost white brothers and he was overjoyed to welcome them.

After Truckee's death in 1860, his son inherited his title, Chief of the Paiute Nation. Known as Chief Winnemucca, his territory covered all of what later became the state of Nevada. Truckee's granddaughter, Sarah Winnemucca Hopkins, attended American schools, wore white people clothes and adopted their customs. Nevertheless, she became disillusioned by the white man's land grabbing, his depletion of the buffalo, the Indians' main source of food, and all the forgotten promises. Loyal to her people, she worked tirelessly for them against the greatest odds and constant betrayal. She even made a trip by horseback to San Francisco to call on the general of the U.S. Army for help and later became an interpreter for the army. So effective were her efforts to help the Paiutes that the government invited her to Washington to meet with the President. Unfortunately, the President's promises were never carried out by the Secretary of the Interior.

On the eve of the wagon train's departure, as the members were making last minute preparations, a serious problem developed. A number of men from the neighboring tribe of Winnemuccas, attracted by all the commotion, came over to see what was going on. Moses Schallenberger discovered he was missing a halter and when he spotted it hanging below the short feather blanket of one of the Indians, he demanded that the Indian return it. Receiving no response, he impulsively tried to grab the halter at which point the Indian drew his bow. Moses ran to the wagon for his rifle and drew a bead on the Indian. Reacting quickly, the younger Martin wrested away the rifle so that it shot up in the air.

The whole camp was in an uproar. Red men and white men gathered around. Moses was filled with remorse for having acted so impulsively. After a strong dressing-down from his stepfather, he felt even worse. Following the wise advice of the elder Martin, the white men apologized and appeased the Indians with gifts of blankets and food. If the Indian had been killed, according to Old Greenwood, there was no doubt that the entire party would have been massacred. The Indians had not yet acquired firearms, but, adept with bow and arrow, Greenwood said they would have been capable of great harm.

CHAPTER FIVE

Conquering the Mighty Sierra Nevada

After the nearly disastrous episode with the Winnemuccas it was urgent that the wagon train move on. At two in the morning, with Old Greenwood in the lead, the emigrants struck out toward the Sierra Nevada Mountains across a sandy desert that was almost totally lacking in vegetation. They were making the first wagon tracks across the northern part of the region that became known as the Forty Mile Desert. Their route would be followed by emigrant trains in succeeding years. Remembering well the ordeal of the Greenwood Cutoff, the Murphy women had prepared two days' cooked rations and vessels filled with water.

From an Andrew Hill painting depicting the Murphy party about to cross the Sierra. COURTESY ANN DERBY JOY COLLECTION.

The deep alkali dust of the desert was especially stifling for Mary Bolger who was now in her eighth month of pregnancy and for Ann Martin Murphy in her sixth month. After twenty-four hours of hard traipsing, they stopped to give the women and the oxen a short rest at what was called Hot Springs Station after the Central Pacific Railroad came through. It was urgent that they keep going so they continued on for another twelve hours.

As soon as they sighted cottonwood trees along a river they knew this was the river Truckee had described. With the Green River experience still fresh in their minds, they unhitched the oxen to prevent their rushing into the water, wrecking the wagons and their contents. In appreciation of Truckee's advice, the elder Martin suggested they name the stream for him. Their worries about water and feed for the animals were over for the present, but unfortunately the animals' thirst was so great they over-indulged. Several became ill, making it necessary to stop for two days.

Although there were no wagon tracks or roads, with pleasant weather and plentiful game they made good progress. After a few days they passed an open meadow which later became the site of the city of Reno. Continuing westward, the mountains started closing in and the stream became so crooked that they sometimes crossed it ten or more times in a mile. The members of the party were fatigued yet they dared not rest. Disregarding her husband's protests, Mary Bolger insisted on doing her share of walking.

By the middle of October light snows began to fall. Fearing they might become buried in the snow, or that they might not be able to reach the pass in time to cross the mountains, Captain Stephens urged the party to press on. Soon the hills moved in so close that with the dense growth of cottonwood and willow there was no longer enough room between the river's edge and the hills for the oxen to get a foothold.

The men were now compelled to travel with the wagons in the bed of the river. The feet of the cattle had become so tender and walking on the rough rocks so painful the drivers had to walk beside them hip deep in the icy cold water, urging them to take each step. At times it was necessary to increase the yokes of oxen to pull the wagons over large boulders and steep rises. To make matters worse, a foot of snow fell, burying the grass out of the oxen's reach.

At night the poor beasts stood and bawled so piteously that the emigrants forgot their own misery. They had nothing to give them but a few pine needles. When at last they came to some rushes peaking above the snow the animals ate so greedily that two of Jim Murphy's cattle died. Each day now, scouts were sent out to find rushes and locate the next night's camp. Onward they pushed knowing there could be no thought of retreating. Winter was at hand as they reached a fork in the river. Once again they were faced with a momentous decision and concluded it would be best to spend the night.

At this site, where the town of Truckee would be born, the main stream, the Truckee, angled south and the tributary, Truckee Creek, west. After much discussion and with some dissension, it was finally agreed they would take Old Greenwood's advice to follow the tributary to the west. To follow the larger stream along the edge of the mountain and through a dense forest would be time-consuming and at times impossible because of the wagons.

As a protective measure it was decided that some of the party should separate from the wagons and go ahead by horseback following the main stream directly to Sutter's Fort. In the event that something should go amiss with the wagon group, they might enlist Captain Sutter's help. Making up a party of six were four men; Dan Murphy, the leader, his brother, John, Olivier Magnent and the Townsends' servant, Francis Deland, each carrying a rifle and ammunition, and two women, Helen Murphy, and Elizabeth Townsend—each had a change of clothes and blankets. With two packhorses to carry provisions and ammunition, they were on their way.

The party with the wagons proceeded up the tributary two and one-half miles until they came to the eastern end of a lake surrounded by pine trees—no doubt they were the first white men to behold its stunning beauty. They named the lake and the pass for their hero, Truckee. Much to Captain Stephen's displeasure, however, both names were later changed to Donner for the party that spent the tragic winter there two years later.

Here they set up camp. The weary emigrants looked across the lake at the ominous sight of the towering, snow-covered granite mountain they must cross. It appeared to be an insurmountable barrier, but they were sustained by the words of Father Hoecken who had so eloquently described to the Murphys the Promised Land they would find on the other side.

After days of searching for a pass, Old Greenwood found one, but it was so narrow that many decided to leave their wagons at the lake. Those determined to get their wagons over the top were the younger and elder Martin Murphys, Jim Murphy, James Miller and Isaac Hitchcock. Moses Schallenberger volunteered to stay at the lake with the remaining wagons until spring to protect Dr. Townsend's valuable merchandise that included brocaded silks and satins which he planned to sell in California. Joseph Foster and Allen Montgomery offered to keep him company after first agreeing to escort the wagon party to the summit in the event they should need help.

The vertical rock in Donner Pass that the Murphy Party had to surmount in order to conquer the Sierra. FROM *TRAIL OF THE FIRST WAGONS OVER THE SIERRA NEVADA* BY CHARLES K. GRAYDON.

In snow now two feet deep the Murphy-Stephens party prepared to ascend the formidable mountain one thousand feet above them. The steep slope was covered with boulders whose slick surface made difficult footing for the oxen, a challenge which was added to that of the wagons having to make their way through the forest of large pines. First they would unload the wagons and, no small task, carry their contents and the small children to the top of the pass. Then with the young men, including Dan and John Murphy, Dennis Martin, Moses Schallenberger and John Sullivan pulling and pushing, the wagons were dragged upward inch by inch. Halfway to the top they came to a ten-foot high vertical stone wall. At this point all appeared to be lost. They feared they would have to abandon their belongings except what the men could carry on their backs.

They were finally saved when, after a frantic search, Old Greenwood found a rift in the wall just wide enough for one ox to pass. After removing their yokes they led the animals one at a time to the top where they were reattached and the chains were lowered to be fastened to the tongues of the wagons below. With the combined determination of the younger men, including Martin, Joseph Foster, John Sullivan, James Miller and others, by lifting and pushing the wagons from below and the oxen pulling from above, the wagons were hoisted over the top. Great was their elation as they rehitched the oxen to the wagons and continued up the slope to the summit.

They had just surmounted the almost insurmountable to become the first wagon train to reach the Sierra Divide, thus opening the California Trail for thousands of emigrants to cross as well as providing a route for the railroad and still later for U.S. Highway 80. The Murphy-Stephens party was the first wagon train to bring wagons and American cattle into California. The date was November 25, 1844.

From the summit, like the Israelites, they looked out on the Promised Land described by Father Hoecken—yet with the urgency to move on in the ever-deepening snow they dared not tarry. Not only was Mary Bolger's baby due any time, but they feared being snowbound. After three days they came to the headwaters of the Yuba River.

The leaders now faced the most difficult decision of the long journey. Realizing it would be folly to try to take the wagons further, the men agreed they had no alternative but to build a camp for the women and children. The spot they chose would later be known as Big Bend. On December 1, 1844, Mary Bolger gave birth to a baby girl—the second delivery of the journey for Dr. Townsend. Much to the younger Martin's relief, he was now able to leave with the knowledge that mother and child were doing well. Little Elizabeth, named for her mother's twin sister, was the first child born in California of American emigrants. She would later acquire the middle name "Yuba." Also staying at the camp on the Yuba were William Miller with his wife, Mary Murphy Miller and their three children, Mary Bolger and her sons, James, Martin, Patrick, Bernard and baby Elizabeth Yuba, and Mrs. Patterson and her five children.

With the completion of the camp, the men now had a soul-searching decision to make. They concluded they would have to leave the five wagons at the camp. Capable James Miller volunteered to stay with the women and children, joined by old Mr. Martin. With mixed emotions the others left for help at Sutter's Fort—some on foot, others on horseback and others driving a half dozen or so cattle to break the way through the snow.

Following the Yuba River, they soon came to a broad valley (Bear Valley) and from there followed the Bear River. Forty miles before reaching Sacramento, they stopped at Johnson's Ranch to which they would be followed by countless gold seekers after 1849. They entered the Sacramento Valley, arriving at John Sutter's Fort at New Helvetia on December 13, 1844.

CHAPTER SIX

Together at Sutter's Fort

After a stop at Johnson's Ranch the wagon party arrived at Sutter's Fort—relieved and excited to find the party of six, which followed the southern route, had arrived a few days earlier. It was a joyous reunion for the elder Martin as he embraced his beloved youngest daughter, Helen, and his sons, Dan and John. Dr. Townsend and Elizabeth were overjoyed to be together again. Each party had worried about the other, and each was full of questions.

Sutter's Fort, 1849. FROM *CALIFORNIA ILLUSTRATED* BY J.M. LETTS. COURTESY CALIFORNIA HISTORY SECTION, CALIFORNIA STATE LIBRARY.

The younger Murphy boys described how they followed the main stream of the Truckee and came to a beautiful mountain lake surrounded by pine trees growing down to the water's edge. This impressive body of water would later be known as Lake Tahoe, and Dan as the first white man to step on its shores. They told of following the Rubicon River to the American River, and then of continuing to its confluence with the Sacramento River. The American, he said, was wider and deeper but just as

crooked as the Truckee, so that they had to crisscross its banks many times.

Elizabeth added that the Indian pony which her husband, Dr. Townsend, had traded for her at Fort Laramie, was an excellent swimmer. "I would ride him across the river," she said, "and then send him back with one of the boys to get Helen." Their really frightening experience, she added, occurred when John rode the pony back over the stream to get a packsaddle. By this time, the poor animal was tired from several crossings. He lost his footing at a place where the icy cold current was running with extreme force. John was dashed against the huge rocks, but managed to grasp an overhanging branch until they rescued him. Continuing with their story, they said before reaching New Helvetia they came to the ranch of some people named Sinclair and accepted their invitation to stay.

While they were absorbed in conversation, Captain John Augustus Sutter came up and greeted them warmly. A handsome, solidly built, rosy-cheeked man with dark sideburns that joined under his chin, he had met and mastered many challenges before arriving in California to rise to absolute ruler of this princely domain on the Sacramento River covering two hundred square miles. Driven from his own country of Switzerland by bad debts, he had abandoned his wife and family, promising to send for them, which he finally did.

Sutter told his guests he had come via a circuitous route that included merchandising on the Santa Fe Trail, traveling overland to Oregon, and finally, when no sailing ship was available directly to Yerba Buena, detouring by way of Hawaii. In 1839 he arrived in California with his eight loyal *Kanakas*. Putting his powers of persuasion to work, he was able to convince Governor Juan Bautista Alvarado, who already had concerns about protecting the wilderness frontier on the Sacramento River, to grant him forty-eight thousand acres with the provision that he would build a settlement.

The Swiss offered to show the new arrivals around his fortress. With considerable pride he pointed to the eighteen-foot high, one and one-half foot thick wall surrounding the fort, saying that members of the Chucumnes tribe had built it of baked adobe brick. At times he would have as many as one thousand Indians working for him. The practical Murphy men wondered if it wasn't difficult to feed that many men. They were soon sorry they asked when they came to an "outdoor dining room" and saw V-shaped wooden troughs filled with atole, or corn mush, from which the Indians were feeding themselves with their hands—a scene repugnant to the Murphys.

As to their question about problems with his Indian workers, Sutter said that after he got wind of a Chucumnes Indian plot to kill him and capture his fort, he and some of his men made a surprise night attack on the tribe at their camp on the Cosumnes River. In the fracas he and his squad killed six warriors. "Since that time," he said, "I have had no trouble, and they have become loyal to me."

Captain John A. Sutter welcomed weary pioneers arriving at the fort after their long trek across the plains. Oil on canvas by William Smith Jewett, 1856. COURTESY THE OAKLAND MUSEUM KAHN COLLECTION.

Continuing the tour, they passed a squadron of natives goose-stepping to the commands in German of one of Sutter's drill sergeants of the Royal Swiss Guards of France. Sutter, who had delusions of military grandeur, colorfully described his supposed career as captain of the Royal Swiss Guards of France—a story he had repeated so many times he had begun to believe it himself.

Moving along he showed them *Casa Grande*, a three room adobe structure that served as his headquarters, the barracks for the Indian troops, a house for guests, a bakery, a mill, a blanket factory and various workshops. In addition to all this, he told them of his pioneering in agriculture, lumbering, tanning, and shipping—that he had a launch that plied the waters between the fort and Yerba Buena, transporting building supplies such as iron for the blacksmith and food. He was all things to many people: administrator, justice of the peace, physician and patriarch.

After all this congeniality Sutter suddenly turned serious. Now came the moment of truth. He told them California was going through a power struggle involving former Governor Juan Bautista Alvarado, General José Castro, Governor Pío Pico and General Mariano Guadalupe Vallejo. He was sorry to have to tell them that foreigners are no longer welcome. Alvarado and Castro have organized a revolution to depose Governor Manuel Micheltorena.

Sutter, an admirer of the distinguished Micheltorena, explained that the governor had failed to win the support of the *Californios*. They resented the Mexican government's imposing on them its own choice of governor, and they were especially bitter about the *cholo* army, whom they called "chicken thieves," sent by the Mexican government. Showing complete disrespect for the *Californios*, the *cholos* supported themselves by theft from the ranchos. Noting the discouraged expression on the faces of the Murphy party men, Sutter told them they had an alternative, "If you will join my army in support of Micheltorena," he said, "you will be able to stay, and the governor will reward you with land."

Both Martins protested. "We left our women and children at the camp in the mountains. We must return to them with supplies."

Sutter explained that an expedition into the mountains at that time, having to make its way through deep snow, was impossible and he assured them their families would survive on the available food until spring. It was a difficult decision over which they suffered for some time. The Murphys, Stephens and Townsend finally agreed that the families were in the capable hands of James Miller and that he would be able to provide food for them. It was a long way to Council Bluffs and they all conceded that, after coming two thousand miles, they should not turn back.

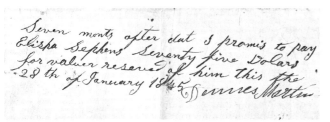

IOU signed by Dennis Martin for Elisha Stephens. COURTESY SANTA CLARA UNIVERSITY ARCHIVES.

Sutter then spoke about the tall, handsome, educated governor and of his plans and accomplishments such as establishing the first public school system in California, regulating medical fees to stop unqualified doctors, putting an end to bribery and smuggling by Mexican officials and returning the missions to the church in the hopes of controlling further deterioration.

The governor, he said, was responsible for turning the tide of Yerba Buena's (San Francisco) fortunes by declaring it a port of entry where foreign ships would pay fair landing fees and duties. He also appropriated eight hundred dollars for a customs building. Until that time Yerba Buena was going nowhere—was mainly an outpost for the Hudson's Bay Co. Foreigners had pilfered Mission Dolores and the *presidio* of adobe bricks and lumber for their buildings.

He told them Micheltorena's troubles started in mid-November 1844 when Manuel de Jesus Castro, with a party of fifty, appropriated horses from the government corrals and drove them from Monterey to Salinas and then to the Laguna Seca Rancho of Juan Alvires about twelve miles south of the *pueblo*. After a confrontation at the Santa Teresa *rancho* of Joaquin Bernal on December 1, between Charles Maria Weber and Micheltorena, the Treaty of Laguna Seca, or as it was also called, the Treaty of Santa Teresa, was signed. In the treaty Micheltorena agreed to send his unruly *cholos*, (a bone of contention with the *Californios*), back to Mexico. He later recanted. Considering the unrest and temper of the times, he believed that if he disbanded his army he would be sawing off the branch on which he was sitting. It was Sutter's contention that, had the Mexican government given Micheltorena one hundred well-trained army men, it might have been a different story.

After the signing of the treaty, Charles Weber, a former employee of Sutter, made a trip to New Helvetia to try to persuade his friend that he was making a mistake in backing Micheltorena. For his effort, Sutter put Weber in the *calabozo*, but soon moved him to comfortable quarters for the duration. As it turned out it wasn't all bad. During his internment Weber met his future wife, Helen Murphy, who, with Elizabeth Townsend, was waiting out the war until her family returned. During the long evenings they became well acquainted.

CHAPTER SEVEN

War, California Style

On the first day of 1845 the sound of a fife and the roll of the drums by two Indians signaled the appearance at the fort's gates of Captain Sutter and his Army of the Sacramento. The fanfare attracted the attention of a scattering of Indians just outside the entrance who watched as Captain Sutter, a proud, handsome figure in his ornate Mexican army uniform, blue frock coat, tilted cap and polished boots, led his men from the compound.

Sutter's shock troops of eighty-five soldiers, commanded by Captain John Gantt, and two American lieutenants, followed. Many of his soldiers were bearded mountain men dressed in buckskins with fringe on their sleeves and sides of their pants. Next came Sutter's Indian troops and his Kanakas from Hawaii. Numbering one hundred, they goose-stepped to the commands of their German captain, Ernest Rufus, whom the Murphys had seen training the day before. He was shouting his orders in English, Spanish and pidgin Mokelumne as well as in his native German. Besides being well drilled, the Indians were impressively dressed in red-trimmed Russian uniforms with flintlocks over their shoulders, items that Sutter had acquired when he purchased Fort Ross from the Russians in 1841. In that deal he also received a brass field piece and two cannon which Napoleon had apparently left behind on his retreat from Russia. An ox cart at the rear carried the Swiss marshal's artillery and supplies.

Sutter's own staff followed the gunners. They included Dr. Townsend, his medical aide de camp; Samuel Hensley, his commissary manager; and John Bidwell, his loyal field secretary who had overseen the transfer of the Russian property from Fort Ross and then managed Hock Farm where Sutter had his home.

The riflemen included Murphy-Stephens wagon train members, fifty-nine year old Martin, sitting tall in his saddle, joined by his five sons, the younger Martin, James, Bernard, John and Dan; Captain Elisha Stephens, Caleb Greenwood and his sons, John and Britain, and Dennis and Patrick Martin, brothers in-law of Jim Murphy.

Other soldiers included the respected mountain man, Peter Lassen for whom the peak in Northern California was named; Kit Carson's brother, Moses; the wild and woolly Ezekiel Merritt who the following year led the Bear Flag Revolt for John Fremont; the notorious Isaac Graham, a Kentuckian who joined Sutter to get back at his enemy, Alvarado; and another man of doubtful character, Grove Cook, who had worked at the distillery at the fort and while there had killed an Indian named Elijah, son of Chief Yellow Serpent, who had come to exchange wild horses for cattle. Also traveling with Sutter were two Indian servants and his Hawaiian courier.

Sutter's plan of attack was to meet and conquer Castro and Alvarado's smaller army of ninety men at Mission Santa Clara. They headed out toward John Marsh's *rancho* at Mt. Diablo, where Sutter, having been misinformed that Marsh wanted to join his army, pushed the issue, forcing Marsh to join him as a private. This later proved to be a costly error on Sutter's part. They continued on past Don Antonio Suñol's *Rancho del Valle* to Mission San José where the padre received them warmly. The *Californios* at the mission, sympathetic to Sutter's cause, were overly hospitable, serving his men more than ample food. Wine flowed so freely that the next day the men were a sad looking lot as they tramped on to the *pueblo* of San José. Sutter encamped outside town, sending word ahead to the *alcalde* that he needed eighty horses and instructions that all the *cantinas* be closed—no spirits that night for this man's army.

They found the *pueblo* nearly deserted and were told by Sutter's friend and sub-prefect of the district, Don Antonio Suñol, who had previously tried to dissuade Sutter from this ill-conceived war, that the *alcalde* had fled and that he, Suñol, had arranged for the *cantinas* to be closed. He iterated what the padre had told Sutter, that Castro was on the run.

Juan Bautista Alvarado. FROM CALIFORNIA HISTORICAL SOCIETY QUARTERLY, SEPTEMBER 1935.

From the *pueblo* they continued down the old Monterey Road, part of El Camino Real, passing through the broad, oak-studded Santa Clara Valley. The mid-day sun shone brightly on the softly contoured Mount Diablo range, now wearing a cloak of brilliant green after the early winter rains. On the western side rose the taller, pine and redwood-covered Santa Cruz Mountains. Entranced by the valley's

beauty the elder Martin told his sons that Father Hoecken was right when he told them this was the proverbial land of milk and honey. Here they were in the middle of winter and the sun was shining, the land was level and the soil rich.

At the end of each day's march the quartermaster, Jasper O'Farrell, laid out the camp. After the pitching of tents necessary orders and the watchword for the night were given. Around the campfire that night, Martin told his sons he now knew where he wanted to settle and that he wanted to build a house looking up at the peak they had just passed, a goal that he would one day achieve. For fifty years travelers knew that peak as Murphy's Peak.

The next day outside the mission village of San Juan Bautista, Governor Micheltorena marched out with his one hundred-fifty *cholos* and fifty military men to join forces with his marshal from New Helvetia. The governor, in the lead, played what turned out to be a delaying game, camping and recamping, always far behind Castro. Yet Sutter was gullible, taken in by the governor's big talk of pursuit and punishment of the enemy. At night they visited each other's camp—Sutter flattered to be in the company of the man he so admired. It never occurred to Sutter that there might be collusion between the governor and the general.

They trailed well behind Castro, who raided the *ranchos* along the way for horses, thus depriving Sutter of fresh mounts for his men. On the canny advice of Alvarado, Castro moved his men on to the *pueblo* of Los Angeles, which they succeeded in occupying. They won the Picos over to the rebel's side as well as poisoned the minds of the *Angelenos* by warning them that Micheltorena and Sutter were going to kill, plunder and burn the *pueblo* to the ground.

It was not the most exciting of wars. The cold, heavy winter rains in the Central Valley created an impossibly muddy situation, slowing the soldiers' pace. The governor became ill with piles and was unable to mount a horse, necessitating a four-wheeled coach to be sent for him. This caused a week's delay and further deterred progress when the coach had to be carried over mountains. They were now only able to travel ten or twelve miles a day. As a result their provisions ran out. The Murphys found themselves living on cattle found along the way and foraging at the missions. The lack of activity—not even a skirmish—made it a boring affair for the red-blooded young Murphys. Facetiously they asked their father if this is what it was like in the Battle of Vinegar Hill in which, he often told them, he had taken part as a teenager in Ireland.

The excitement picked up when some of Sutter's men, on a scouting expedition between Gilroy and Watsonville, took as prisoner rebel leader Manuel Castro who was camped on *La Brea Rancho* near Gilroy. The soldiers were outbluffed a short time later by Charles Brown who recaptured the prisoner. This was the same Castro who had started the whole thing when he seized the governor's herd of horses at Monterey, which prompted the signing of the Treaty of Laguna Seca.

The Murphys were becoming increasingly skeptical. They suspected political duplicity on the part of Micheltorena and General Castro. Just below Monterey the elder and younger Martins, joined by Captain Stephens, talked with Micheltorena about their concern for the women and children at the camp in the mountains and requested permission to take leave to rescue them. The governor granted their request and gave them his blessing. The three men risked attack by the *Californios* who resented Micheltorena's *cholos* for having raided their ranches, but they returned to Sutter's Fort without incident.

Apparently Sutter had some misgivings about having talked the Murphys into joining him because in a letter to Pierson Reading, who was in charge at the fort, he wrote, "I hope the other party of Captain Stevens will have returned safely from the mountain." Sutter had come to know and like the two Martin Murphys. In a letter to Reading he wrote, "This morning I received a courier from Captain William R. Richardson. He fears some danger will be for the establishment (Sutter's Fort). You have watchful friends on the other side, Captain Richardson and Mr. Murphy."

At Santa Barbara, where they spent several days, Sutter, discouraged by their snail-pace progress, decided he had best approach Micheltorena about the grant he was promised. He told the governor of the debts he had incurred for equipment and maintenance of his army. On February 5, Micheltorena gave Sutter the *Sobrante* land grant adjoining New Helvetia, which meant that Sutter's domain now encompassed thirty-three square leagues or two hundred twenty square miles. It was the only positive note for Sutter in the dreary campaign. He sensed the morale of his men was heading downhill. He not only was aware that Marsh was making trouble for him, but he also suspected others, including Gantt, Hensley and O'Farrell.

Although John and Dan Murphy played no part in the dissension, they soon found themselves in a compromising position. While on a reconnaissance mission to San Buenaventura they were taken prisoner along with twelve others by the ubiquitous Lieutenant Manuel Castro. He immediately sent for General José Castro. An American physician named William Streeter, who happened into the camp on his way to Santa Barbara, translated for General Castro. He said the general requested that the Murphy boys give up their arms, adding that it was never the general's intention to drive all the foreigners from the country—that it was Sutter propaganda.

General José Castro. With the former governor of California, Juan Bautista Alvarado, Castro plotted the overthrow of Governor Micheltorena. This led to the Micheltorena Rebellion. FROM *CALIFORNIA HISTORICAL SOCIETY QUARTERLY,* SEPTEMBER 1935.

A few days later Castro offered to release the captives if they would promise not to bear arms against the rebels and would try to persuade the others likewise. Streeter accompanied the detainees back to Sutter's camp, ostensibly for the purpose of discouraging the Swiss from attacking the rebels. John and Dan, together with the other prisoners, informed Sutter they could not continue on, because they had promised they would not participate further in the war.

Streeter continued on to Santa Barbara where he later wrote of seeing the released prisoners passing through and that one of the Murphys (John) was seated on his horse "with a blanket wrapped around him in place of his pants..." and that the other Murphy (Dan) had gone to Micheltorena requesting a pair of shoes, because the ones he was wearing had worn thin from the long walk across the country. The governor told him he'd have to wait until they got to Los Angeles.

The older Murphy brothers, Jim and Bernard, stayed on with Sutter and participated in his one glowing day in the sun at San Buenaventura. In spite of his army having been reduced to less than half, and the interminable winter rain which caused the horses and men to fall on the slippery hillsides, Sutter seized an opportunity for a surprise attack. He learned there had been a *fandango* in town the night before, and figuring correctly that the men were probably still half-drunk and half-asleep, he made a charge on the town. He easily routed them and took possession of the *pueblo*. Now it was his men's turn to get drunk. The same band that had played for the *fandango* and the same Indians from *Rancho San Antonio* with their long black hair and short white skirts, who had danced the night before, entertained them. After this brief, exhilarating moment of victory for Sutter, he was eager to keep the momentum going, but to his dismay, Micheltorena, who had stayed behind, turned him down, saying that it would be better to wait until they could continue on together.

A few days later, as they passed *Rancho Santa Clara*, a beautiful woman, believed to be Alvarado's ex-mistress, came out with her son and inquired where the captain was. When Gantt was pointed out, her son handed him a package of tortillas. Inside he found a letter from Castro and Alvarado calling upon Sutter and his men to abandon Micheltorena and to go over to the insurgents.

The diminished combined forces of the governor and Sutter approached the *Cahuenga Rancho*, part of the San Fernando mission properties north of the pueblo of Los Angeles, where they camped for the night. From a nearby hill they could see the Castro camp. Alvarado sent a messenger under a white flag with a challenge to fight. Micheltorena said he would reply in the morning with the point of his bayonet. A blustery wind raised havoc, blowing down their tents. Yet at dawn the next day, led by fife and drummers, Micheltorena and Sutter marched with their four cannon, as compared to the Castro-Alvarado forces' three, to meet on the southern edge of the San Fernando Valley.

They opened fire at each other from long range while the residents of the *pueblo* looked on in anguish from a hilltop, the women praying their rosaries for husbands, sons, brothers and fathers. The governor's men, under instructions, fired over the enemy's head and the rebels shooting the cannon fled. This was war, California style. Firearms were ruled out as too dangerous. Total losses amounted to two horses on the *Californios'* side and one mule on the Sutter-Micheltorena side.

Micheltorena recognized the tides of war turning against him. Committing a serious blunder he ordered Sutter to send a vanguard of Americans into a gulch within rifle shot of the enemy. With Gantt at the front, the soldiers discovered the men they were opposing were Americans, some of them friends from Missouri. This was the opening John Marsh, who sought revenge on Sutter, had been looking for. Using a wily tactic, with smooth talk he convinced the men they should meet their fellow Americans. It seemed the natural thing to do and thoughts of war were forgotten as they exchanged information about mutual friends. Soon the attitude of Gantt's riflemen was to let Micheltorena's men and the *Californios* fight it out among themselves.

On February 20 Sutter's military career came to an ignominious end. On his way to meet Micheltorena he was surrounded by thirty of the rebels. He feared for his life. Had the *Angelenos*, who considered Sutter their real enemy, recognized him they would have shot him on sight. He was saved by the arrival of Antonio Castro, who told the soldiers he would take care of the prisoner. He sent for Juan Alvarado and, in typical Mexican fashion, Sutter was greeted with *abrazos* from Alvarado. He poured him a drink of *aguardiente* and then called for General José Castro.

The three men accompanied Sutter to the *Cahuenga adobe* where they incarcerated him in a small dark room. After he promised not to escape and with American John Rowland, an early settler, vouching for him, he was taken to the *pueblo* of Los Angeles twelve miles away to be quartered in the house of one of its distinguished citizens, Don Abel Stearns. There Sutter wrote an abject letter of apology to newly appointed Governor Pico who not only pardoned him, but allowed him to keep his fort and empire and confirmed the grant by Micheltorena of three additional square leagues.

The next day Micheltorena, who had little spirit for a fight and realizing it was all over, raised a white flag and suggested a conference. He was permitted to dictate the terms of the *Treaty de Campo de San Fernando* in order to make the best possible impression on Mexico City. Furthermore, he was allowed to march his *cholo* army through Los Angeles, accompanied by music and colors flying, to a camp at Palos Verdes (near San Pedro) before sailing with them back to San Blas. Micheltorena apparently anticipated the U.S. government would be taking over and was glad to be out of it. The likely double-dealing of the governor and the general, unknown even to the new governor, Pío Pico, was now over.

As for Sutter, he was called to defend his conduct before the heads of the new regime. After those gathered had consumed numerous aguardientes, Sutter took an oath of allegiance and was reinstated and redecorated with his fancy titles. Still it was a humiliating end to what was to have been for him a glorious military career—a battlefield of lost dreams. Despite all the fine talk, the *Californios* gave him no assistance for the trip home. A German cooper and former employee came to his rescue and offered him thirty horses on credit. The government allowed him to receive from the San Fernando mission a meal of mush for the Indians, but on their way north Sutter and his men depended for food on the horseflesh of the wild *manadas* of the Tulare valley, fish from the streams and an occasional pronghorn.

When he finally returned to his fort Sutter found that during his three-month absence his home force had done practically nothing. He faced the loss of his crops and one hundred horses. Even the Indians had reverted to their old ways, attacking the *ranchos* of his neighbors.

CHAPTER EIGHT

Snowbound on the Yuba

After leaving Micheltorena's army with the governor's blessing, the two Martins and Captain Stephens, aware of the *Californios'* animosity toward foreigners and especially toward members of Micheltorena's army, passed cautiously through hostile territory to arrive without incident at Sutter's Fort. Pierson Reading, whom Sutter had left in charge of the fort, brought them up-to-date on the rebellion. They learned for the first time of Micheltorena's surrender and of Sutter's being taken prisoner. Reading also told them that the snows in the mountains had been unusually heavy. He regretted that he couldn't give them very much food because of the loss of wheat crops, but said that he would send more provisions with Sam Neal, a blacksmith at the fort.

Meanwhile, at the camp on the Yuba, worries persisted regarding the men's long absence and the critical food shortage. James Miller, who was caring for the needs of the women and children, decided to set forth with his twelve year-old son, William, to try to find the missing men as well as to locate food. After almost a week of trudging through snow, William's face lit up with a sign of hope. "Father," he said, "I hear bells. It sounds like pack animals."

James listened closely. "For sure," he said, "it must be a relief party." Soon the Murphys and Captain Stephens came in sight. With great relief James told them of the predicament at the camp—that they had become desperate for food. "We have even boiled shoes to try to ease the pain of hunger," he said.

Shocked, the younger Martin said, "We had no idea you were in such a dire straits. We were so eager to see you all that we brought rather scant provisions, but Pierson Reading is sending horses and more food."

When they arrived at the cabin the families forgot their hunger for the moment as they rejoiced with much hugging and kissing—questions flying back and forth. Starting at the beginning, the younger Martin told them of joining Sutter's army—explaining that they really had no choice—their only alternative was to turn around and go back to Council Bluffs. Everyone agreed they wouldn't have wanted to do that.

Mary Bolger had disturbing news about Moses Schallenberger, who had almost become a member of the family. She remembered aloud how he had carried her little Barney on his shoulders halfway across the country. Joseph Foster and Allen Montgomery had come by to get Allen's wife, Sarah. While at the cabin they talked about their experiences after leaving them at the summit and traveling on with Moses. The first thing they did when they got back to the lake was to build a cabin. The morning after they completed it, to their surprise, they awoke to find the snow had fallen to a depth of three feet and that it continued to fall so that they were unable to go out after game.

"At first we were not concerned," Foster said, "because we expected the snow to melt. But instead it continued to fall." They killed the two skinny cows they had been given—rationalizing that the animals would have starved anyhow because they could not reach food that was covered with snow.

After consuming half their meat, including the dried buffalo they had brought with them, the men finally faced up to the fact that the three of them would have to try to make it to California on foot or face starvation. Never having had experience in deep snow, they did the best they could by improvising snowshoes of wagon bows filled with webbing of rawhide. Off they started—each with a pair of blankets, their rations of beef, and ammunition.

Montgomery said that the weight of lifting their makeshift snowshoes was especially hard on young Moses, for he was still immature. His legs cramped and he became so ill that finally he insisted they go on without him and that he would make his way back to the cabin.

Foster, with concern in his voice, said that the snow was coming down so hard they hoped he had made it. After finishing their story, Foster and Montgomery, joined by Sarah Montgomery, continued on their way to New Helvetia.

The day following the two Murphys' return to the cabin, Moses suddenly appeared there with Dennis Martin. Moses told of staggering back to the cabin after he left Foster and Montgomery and of his despair that he would ever be able to leave. Turning to Stephens, he said, "Using the traps you left I survived on fox I was able to trap. I found the coyote meat horrible, but saved a supply of it as a last resort." The time there seemed interminable. His only relief was to read the books from Dr. Townsend's extensive library.

Putting his arm around Dennis, he said, "Then one day I looked out and saw a figure coming toward the cabin. It was Dennis. He was a true hero to find his way through that deep snow with no paths and no landmarks."

Dennis explained that he met Moses' sister, Elizabeth Townsend when he arrived at Sutter's Fort. She begged him to try to find Moses. He promised her he would. "Having lived in Canada," Dennis said, "I was familiar with ways of getting through the snow. I made a more practical pair of snowshoes for Moses so that he wouldn't have to lift up the whole shoe with each step."

To add to the "old home week" atmosphere, a few days later John and Dan arrived at the cabin. They, too, had a story to tell. John described how, when they were on a reconnaissance mission at Buenaventura they were captured by Castro's men. They were released a few days later after promising not to fight against the rebels. They said they were sick of the war and of Micheltorena's indecision and seeming lack of desire for victory. "I suspect," Dan said, "there must have been a pact between the governor and Castro, and poor Sutter never suspected."

When Sam Neal arrived with provisions and horses from the fort it was a time for celebration. It had been months since leaving Council Bluffs. Now the end seemed to be in sight. But their optimism would be short-lived.

The band of emigrants set off the next day, some with their wagons and others mounted, eventually making camp on the banks of the Bear River. It was the first of March. The river was full and still rising from the melting snow in the mountains and the heaviest rainfall of the season. It rained hard all that night and the next morning. They could find no bridge or ferry to cross the river.

Sam Neal, who had gone ahead down the stream on a search mission, was cut off from the shore by rapidly rising waters. He found himself on a little island that was losing ground to the flood. Unable to swim, he climbed a tree. His cries for help soon reached the ears of the others. To the rescue came Moses Schallenberger and John Murphy. They immediately mounted their horses and, leading a third, swam into the swirling torrent to bring him safely to shore.

What had started out as a joyous occasion turned somber. The supply of provisions was exhausted. The Bear River was now ten miles wide. Dan Murphy was able to bag two small deer—scarcely a mouthful or two for each member of the large party of adults and children. At this point there was no turning back. Adding to their anguish, they could hear the lowing of the cattle across the river and occasionally caught sight of the graceful herds of pronghorn…so near and yet so far.

They finally met a herd of wild horses ranging on the hillsides, but the younger Martin Murphy could not tolerate the thought of eating horsemeat. They chose a time when he was out looking for an ox that had wandered away on their trip to Sutter's Fort. Schallenberger shot a fine, fat two-year-old filly. When Martin returned it was dressed and roasting on the fire. He was delighted and commended Moses, exclaiming, "Good boy, good boy, but for you we might all have starved." He ate heartily, declaring it to be the finest meat he ever tasted. His amused brother-in-law, James Miller, couldn't resist telling him what he had been eating. The younger Martin's sensitive stomach rebelled and he lost the meal he had enjoyed so much.

Soon after this incident the waters receded sufficiently to allow them to reach the Feather River near Sutter's residence, Hock Farm. Sutter's foreman, John Bidwell, had a boat prepared to ferry them across. A near-catastrophe was averted when the saddlebag, in which Elizabeth was secured to the pummel of her mother's saddle, became loose. She fell into the river, but was scooped up quickly by her father. In nervous relief, he proclaimed, "Your middle name shall be Yuba." The name stuck, and she was forever after called Lizzie Yuba. Perhaps this experience brought her closer to him, because the family always said she was the favorite of his three daughters.

Hock Farm, the residence of Captain Sutter on the Feather River. COURTESY CALIFORNIA HISTORY SECTION, CALIFORNIA STATE LIBRARY.

John Bidwell, a loyal employee who managed Hock Farm for Captain Sutter. COURTESY CALIFORNIA HISTORY SECTION, CALIFORNIA STATE LIBRARY.

Elizabeth Yuba Murphy, who was born at the camp on the Yuba River. COURTESY SUNNYVALE HISTORICAL SOCIETY AND MUSEUM ASSOCIATION.

As Martin and Mary Bolger approached Sutter's Fort with the four boys in back of the wagon and baby Lizzie Yuba in her mother's arms, they gave thanks to the Almighty Father for their blessings—for their safe sailing of the Atlantic from Ireland to Canada, for the crossing of the rivers, lakes and canals to Missouri, for the long, arduous journey across the seemingly endless plains, and finally, for the conquering of the mighty Sierra Nevada.

Now, with all that behind them, they were on their way again. But Martin and Mary had big decisions to make. They rode along, deep in thought, to the top of a hill. As they looked down on the broad, level Sacramento Valley, freshly green from the spring rains, they agreed this must have been the way the Israelites felt when they saw the Promised Land. Martin, in his soft Irish brogue asked, "Mary, my darlin', where in this land of plenty as Father Hoecken told us it would be, do you think we should settle?"

Mary's usually animated face was pensive. Finally, she wondered aloud if they should go any farther. Martin smiled and said he had been thinking the same thing. Always the farmer and always practical, he talked about their growing family that must be fed. If they stopped here there would be good soil for wheat and ample feed for cattle. Besides, he added, it would certainly please Sutter because it was part of the agreement with Governor Alvarado that a settlement would be built on this forty-eight thousand-acre grant.

At Sutter's Fort the Murphy-Stephens party gathered together for the last time. They talked of their plans for the future. John and Elizabeth Townsend said they were staying on at New Helvetia, as were Allen and Sarah Montgomery and Edmund Bray. The elder Martin said he was going to the *pueblo* of San José with his four unmarried children to see about buying the *rancho* land they had passed on their way south with Sutter's army. The younger Martin said that he and Mary had decided to stay in the valley with their children to farm—that he planned to apply for Mexican citizenship in order to buy a *rancho*. James and Mary Miller would be going on to Mission San Rafael and hoped to farm near there.

CHAPTER NINE

The Millers of Marin

After their snowbound winter in the cabin on the Yuba River, James and Mary Murphy Miller found Sutter's Fort a welcome respite. Although Sutter had not yet returned from the Micheltorena Rebellion, while at the fort they met another man, one who would steer them and their descendants on a separate course.

In a conversation with Yerba Buena Port Captain William Richardson, Miller asked about California and where he would suggest they should settle. Richardson, grantee of the nineteen thousand-acre Rancho Saucelito (where the city of Sausalito would one day stand) told James and Mary about the land adjacent to his. He said it was located near Mission San Rafael in the shelter of Mount Tamalpais, that it had a mild climate and fertile soil. His enthusiastic description convinced the Millers. James and Mary decided right then and there that San Rafael it would be.

James Miller stayed with the women and children at the camp on the Yuba River while the other men went to war. COURTESY MARIN COUNTY HISTORICAL SOCIETY.

Mary Murphy Miller. COURTESY RUTH MURPHY POLK COLLECTION.

Filled with excitement over the prospects of finding a permanent place to build a home, Mary tried to persuade her father to join her and her unmarried brothers, Dan, John, Bernard and her youngest sister, Helen. But her pleading was to no avail. Even though the patriarch hated to be separated from Mary and James and their family, when he rode south with Sutter's army in the Micheltorena Rebellion, he had fallen in love with the Santa Clara Valley and was not about to change his mind.

Mary was sad to be parting from her family, but the Millers' fate was sealed. James lost no time loading all their earthly possessions in the wagon. With William, Catherine and Mary in the back, and little Nellie Independence, now six months old, sitting on her mother's lap in front, they were on the road again. As visions of this new land, which Richardson had so enthusiastically described, danced in their heads, they followed a route alone the shoreline of San Pablo Bay to become the first pioneers to reach overland what later would be called Marin County.

The next morning the sun shined brightly from an azure sky as James mounted his mare to take off on a scouting expedition through the redwood-forested hills. His excitement grew as he crossed streams flowing with crystal-clear water, green meadows, rolling hills and wide valleys. His eyes feasted upon herds of wild horses and longhorn cattle numbering in the thousands—just as Richardson had described. The ordeal of that frozen winter on the Yuba was all but forgotten. He returned to his family in a high state as he described to Mary all that he had seen.

"My dear," he said, "I don't think we need look any further. With all that wild livestock and the rich soil, I could combine farming with cattle raising."

The always amiable Mary, who trusted her husband's judgment, said, "James, that sounds fine. Whatever you say, it shall be." The Millers quickly adapted to Spanish California and, even though there were few Californios in the area at the time, they were soon speaking Spanish and meeting new friends.

On another day of exploring, James came upon Mission San Rafael where he met its major-domo, Timothy Murphy. Better known around the countryside as Don Timoteo, he had become a Mexican citizen after coming to California in 1828. The two men were pleasantly surprised to learn they were both from villages along the River Slaney in County Wexford and that he and Mary Murphy might even be related. An imposing figure—the blue-eyed Don Timoteo stood well over six feet tall and weighed three hundred pounds. To the smaller-of-stature Indians he may have seemed like a giant. With his genial manner, speaking Spanish with an Irish brogue, he endeared himself to the Californios.

Following the secularization of the missions, Don Timoteo taught the Indians to use the agricultural experience they had acquired working on the ranchos.

Don Timoteo was a close friend of Mariano Vallejo. Encouraged by the general, he fell in love with his sister, Dõna Rosario. Even though he sought her hand with the general's blessing, she turned him down for an American from the Middle West named Jacob Leese. Apparently she was Don Timoteo's one true love, for he never married.

A genial host, he built a two-story adobe house in which he entertained graciously and generously with large parties. After his death in 1853 his adobe became the courthouse until 1873. Lieutenant William Tecumseh Sherman, en route to Sutter's Fort with his men, spent a night at Don Timoteo's adobe, and Lieutenant John Charles Fremont, on his way from Oregon with his one hundred-thirty men to take part in the Mexican War, made his headquarters at Don Timoteo's home. Fremont's men camped for several days at Mission San Rafael, providing a little excitement for the Miller boys who were fascinated with the Delaware Indians, part of Fremont's California Battalion.

To assist Don Timoteo in running his large estate, including his twenty-two thousand-acre land grant, the Santa Margarita, the San Pedro and Las Gallinas, he brought his nephews John and Matthew Lucas over from County Wexford. On one occasion, when the elder Martin rode his pony to San Rafael to visit with his daughter, Mary, and the Miller family, he was delighted to meet the young men whose father, Harry Lucas, had been an acclaimed hero in the Irish Rebellion of 1798 in which the elder Martin had also taken part.

Eager for news of Ireland, Martin's eyes filled with tears when they told him about the potato famine that desolated his homeland, known as Ireland's "Great Hunger." They told him millions of people had lost their lives needlessly—that the British were producing plenty of food and selling it abroad. In a quavering voice he murmured, "Oh, my unhappy country, will your suffering and sorrow never end?"

At Don Timoteo's death he willed his land in San Francisco (on which the Palace Hotel was later built) and 680 acres of the Mission San Rafael property to

Bishop Alemany for St. Vincent's Orphanage and School. He left the rest of his property to his nephews. On Matthew's property was developed McNear Beach, Peacock Gap and China Camp. Lucas Valley was named for John.

Don Timoteo had already sold Miller six hundred acres of *Las Gallinas Rancho*, the same land on which the Millers had camped their first night in Marin. Theirs was the first grant deed written in English in Marin County. It stated succinctly: Granted, Bargained, Sold, Aliened, Remised, Conveyed, Confirmed and Released.

In 1849, following the successful path of the elder and younger Martins and with the assistance of his son, William, and his *vaqueros*, James drove one hundred head of cattle to the foothills of the Sierra at Placerville and sold his stock for a neat profit to merchants who were providing meat to the prospectors.

James let no grass grow under his feet. He would eventually become one of Marin's largest landowners, holding eight thousand acres. The following year he acquired the rest of the rancho for fifty thousand dollars. In addition he owned considerable property in the growing town of San Rafael. He first built a Shaker house for his family, then an adobe. Finally, in 1853 as his fortunes soared in farming, breeding cows, horses and mules and especially in an ever-growing dairy business, he built Miller Hall, an impressive English manor house of redwood from trees that grew abundantly nearby. Surrounded by hundreds of acres of park-like gardens, the house was located four miles north of Mission San Rafael on the Petaluma Road, across from St. Vincent's Orphanage. It accommodated his large family, now numbering seven daughters and three sons, who filled the great house with joy and laughter. Hospitable Mary Miller, in the Murphy family tradition, welcomed one and all including her children's friends, so that it became a popular gathering place—especially for the young people.

Miller pioneered education in Marin by building its first school. A mission padre, Father Robetti, was the teacher. Non-sectarian in his educational thinking Miller donated land for the Dixie Public School, Marin's first, built along the banks of Miller Creek. The older Miller children had tutors and governesses, after which the three sons went on to Santa Clara College.

Miller Hall. As the family grew to ten children, James Miller built a spacious home to accommodate the family and their children's many friends. COURTESY MARIN COUNTY HISTORICAL SOCIETY.

Ellen Independence Miller born on the Fourth of July at Independence Rock. COURTESY MARIN COUNTY HISTORICAL SOCIETY

The seven Miller daughters attended Notre Dame Academy in San Jose where they were much loved by the sisters. Ellen Independence, born on the Fourth of July at Independence Rock, a stopping place during the crossing of the plains, grew to be a great beauty and socially prominent. Josephine married Joseph Leonard Kirk, one of the leading lawyers of the state and attorney for the Board of Trade of San Francisco. Teresa married James Ross, for whom the town of Ross was named. The most talented of the girls, Frances (always called Fannie), made Miller Hall famous for her entertainments and popularity as a speaker. Well-known was her art collection acquired during her travels in Europe and her poetry, which was influenced by her good friend, Ina Coolbrith, poet laureate of California.

When B.D. (Barney Murphy's) wife, Annie McGeoghenan, took their son, Martin, east to enroll him in college at Georgetown in 1870, she invited Fannie, her daughter, Evelyn, and her niece, Maud Arques, whose son, Irvin, became a well-known attorney in San Jose, to accompany them on the train trip. In a little book Frances wrote about the trip, "Snapnotes from the Diary of Fannie de C. Miller," she described riding along the Humboldt River traveled by the Murphy-Stephens party in 1844.

As they headed towards Elko she noted, "Passing several small stations we come upon Halleck which embraces four houses on the south side of the track and Uncle Dan Murphy's large dwelling standing alone on the north side." Referring to his separation from his wife, Mary Fisher, she commented, "as isolated of cheerful surroundings as is a man's life in the midst of a divided household." Talented Fanny Miller's life was cut short when she became ill on a trip to Rome where she had gone for an audience with the pope.

Before depositing young Martin at Georgetown University, the travelers went on to Baltimore where they were entertained by relatives of their grandmother, Mary Foley Murphy, who had been a victim of malaria during the Murphys' stay at Irish Grove on the Missouri River. The Foleys became successful in business and named their country estate Enniscorthy for their birthplace in County Wexford. Two of the brothers were anointed archbishops of the Catholic Church—one in Chicago and the other in Baltimore.

The Millers' eldest son, William James, learned a lot about life early. He was twelve years old when they crossed the plains and spent a torturous frozen winter on the Yuba. After his graduation from Santa Clara College he moved to San Rafael where he soon proved that he had inherited his father's drive and ambition. He first built the Marin Hotel and then an entire business block. After Governor Booth signed the bill for the incorporation of San Rafael in 1874, the citizens, numbering about six hundred, elected William one of its first five trustees. With Murphy blood in his veins, he couldn't resist a fling at politics. In 1869 he was elected to serve in the state assembly.

Mary Murphy took ill in 1881. To her regret she could not be among the thousands of guests attending the grand and glorious 50th wedding celebration at Bay View of her brother, Martin, and his wife, Mary Bolger. The gentle, much-loved Mary Miller died the following year. James survived her by eight years—his death coming in 1890. Appropriately, the services for this man, so dedicated to the welfare of orphan boys, took place at St. Vincent's Orphanage across from Miller Hall.

CHAPTER TEN
Settled on the Cosumnes

As soon as John Sutter returned from his detainment in Los Angeles, the younger Martin asked him about buying land to farm and on which to raise cattle at or near New Helvetia. The captain encouraged him to stay. Not only did he like this young Irishman, but also he was eager for more settlers nearby. He offered Martin a job at the fort—said he would help get his Mexican citizenship from General Castro—that the general had told him he would honor Micheltorena's promise to help those who served in his army procure land.

Both Sutter and Castro kept their word and in May of 1845 Martin purchased two leagues of *Los Cazadores* land grant (always called *Rancho Ernesto*) for two hundred-fifty dollars from Ernest Rufus, the German officer who was in charge of Sutter's Indian dragoons during the rebellion. Because Martin couldn't sign his name, the deed was written in the name of his eldest son, James. The Ernesto, located eighteen miles from New Helvetia near the present city of Stockton, covered two square leagues of flat land along the north side of the Cosumnes River.

Pioneering virgin land was nothing new to Martin and Mary after Canada and Missouri, but building an adobe house was a first. They accomplished this with the help of Indian labor sent by Captain Sutter and the older Murphy boys, twelve year old James and nine year old Martin.

There was a lot to interest the younger boys, Patrick and Barney, as they cantered their horses over the valley, entranced by the herds of pronghorn, deer and wild horses—sometimes as many as 500 to 1,000 animals loping gracefully across the plain. At the same time they were always watchful for animals such as bear and longhorn cattle. They learned to shoot some of the birds that haunted the tule marshes—ducks, cranes, pelicans and, at times, thousands of wild geese. They fished in the lakes that formed near the house. When the bottomlands near the house overflowed, fish from the river came up by the thousands—a young boy's dream come true.

The Murphys, having experienced generally good relationships with the Indians during their crossing of the plains, welcomed the local tribes when they came on their rafts, made of pine logs tied together with grape withes, to fish in the Ernesto's lakes and streams and to hold their war dances. When Chief Ole Secumnes married an Indian maiden named Pamela at a nearby *rancheria*, everyone joined in the celebration. At harvest time they came to cut the grain with sickles and, after being collected in circle stacks, it was trampled by a band of wild mares and thrown against the wind to separate the chaff. The grain then fell uninjured onto a sheet.

Their friendship with the Indians would serve the Murphys well when a tribal war occurred the year after they arrived. Even though the chief had just been killed, his widow, out of loyalty, lay at the threshold of the door leading into the Murphy house and another Indian woman positioned herself at the rear for Mary Bolger's safety.

The Indians' festivals fascinated the Murphy boys as they watched the men, dressed in feathers covering their heads and loins, squatting and vaulting, pounding the earthen floor with naked feet as they performed their acrobatic gyrations. All the while they kept time to the beating of drums and the singing of the women as they swayed back and forth in rhythmical motion. Another ceremony that intrigued them featured the medicine man as he sat atop the sweat house for hours haranguing the Great Spirit in a loud monotone, invoking the blessing of good harvests of penola (wild wheat), acorns, game and fish.

Not only did the Indians come to the Murphys' lonely outpost, but Hudson's Bay Company trappers and other figures in the history of that period stopped on their way to Sutter's Fort from Yerba Buena or to the *Pueblo de* San José. Lieutenant William Tecumseh Sherman, later a Civil War general remembered for his march through Georgia, was an overnight guest as were prominent authors of the period including Edwin Bryant and Bayard Taylor who wrote about their visits.

John C. Fremont, who bought horses and cattle from Martin, made the Ernesto his home base of supply and preparations while his soldiers rested and he surveyed the surrounding territory. He explained his presence to Martin saying that he was leading his third expedition west and that his orders were to find a route to the Pacific. Among those in his party were his guide, Kit Carson, Alex Godey, his artist and six Delaware Indians.

The Murphys were a close-knit family and seldom separated for long. Martin's brother James and his sister Mary's husband, James Miller, soon stopped by from San Rafael on their way to Sutter's Fort to get equipment for cutting lumber in the hills. Other visitors included Joseph Foster, who accompanied Martin to Sutter's Fort, and Dennis Martin, who had become a hero when he rescued Moses Schallenberger at Donner Lake. Dennis was on his way to cut lumber and to arrange with Sutter for Indian laborers. He told Martin he had bought parcels of the *Rancho Canada de Raymundo* and the *Corte de Madera*, (part of what is now Woodside) midway between the Pueblo de San José and Yerba Buena. Dennis, a devout Catholic, complained that there were no churches between Mission Santa Clara and Mission San Francisco de Asís at Yerba Buena (better known as Mission Dolores) and that he was planning to build a chapel on his land.

William Tecumseh Sherman stopped at the Ernesto looking for the schoolteacher Martin, Jr. had hired to teach his children. Apparently the teacher was an American soldier who had gone AWOL. COURTESY MARJORIE PIERCE, EAST OF THE GABILANS, VALLEY PUBLISHERS.

Kit Carson and John Charles Fremont stayed at Murphy's ranch on the Cosumnes River. Fremont ordered the appropriation of a remuda of horses stabled in Martin, Jr.'s corral by a Lieutenant Arce to be delivered to General Castro at Mission Santa Clara. This was considered the first act of the Mexican War. COURTESY DENVER HISTORICAL MUSEUM.

Shortly after getting settled on the *rancho*, the younger Martin, ever concerned for his children's education (in fact it was his top priority) located a teacher named Patrick O'Brien. He told his wife, Mary, that the man had apparently suffered financial reverses, signed on with the army and crossed with Fremont. He suspected O'Brien was a deserter, but, in the West he told her, one doesn't ask questions. Besides, teachers were hard to find and this man was well educated. Thus was born at the Ernesto the first school in the Sacramento Valley—one similar to a hedge school in Ireland. O'Brien, a stern disciplinarian, frightened four-year-old Barney (Bernard D.) so much, he said in a Bancroft Library dictation, that for months he couldn't learn his letters.

The children's studies were interrupted from time to time when O'Brien, who had a weakness for alcohol, went AWOL. When he was ready he would show up as though nothing had happened. The boys were amused with his love for a fight. His tenure as teacher came to a temporary halt one day when William Tecumseh Sherman visited the ranch, arrested O'Brien and took him away. Martin later persuaded the young officer to permit the teacher to return, pleading that it was the only school in the valley and that children from other families attended as well.

On one of his regular trips to Sutter's Fort, Martin met the Patrick Breen family, members of the tragic Donner Party. They had just come down from the mountain—wan and weak from their frozen winter of despair at Donner Lake on the site where the Murphy-Stephens party had camped two years earlier and where the Breens and their eight children stayed in Moses Schallenberger's cabin. Martin invited them to come recuperate at the Ernesto. He later accompanied Breen and two other men to the lake to try to recover the family's wagons and personal possessions.

Margaret Breen and Mary shared much in common including their maiden name Bolger. Margaret was from County Wicklow which bordered County Wexford, and both were married in Canada before coming to the United States. The Breens stayed with the Murphys for several months to regain their strength before setting forth for an unknown destination. They eventually found a permanent home at San Juan Bautista where they were the first Americans to settle in that mission community.

Padre Anzar had first taken them in at the mission and after their eldest son, John, made a successful trip to the gold fields, they were able to buy General José Castro's adobe on the *plaza* opposite Mission San Juan Bautista. It was in this house that Castro and Juan Bautista Alvarado plotted the overthrow of Governor Micheltorena. For a while, because of the influx of travelers between Monterey and San José and the need for accommodations, Margaret Breen ran an inn.

On June 10, 1846 the first act of hostility by the Americans leading to the Mexican War occurred at the younger Martin's *rancho*. A Mexican lieutenant named Francisco Arce, driving a *caballada* of one hundred-seventy horses from Sonoma to be delivered to General Castro at Mission Santa Clara, stopped for the night at the Ernesto. Seven Americans, under orders from Fremont, made a surprise night attack. They were about to abscond with the horses when Martin and Mary Bolger, hearing all the commotion, came out and recognized the leader, Captain Ezekiel "Stuttering" Merritt, an officer in Sutter's army during the Micheltorena Campaign. A rough, tough Rocky Mountain trapper, who dressed in buckskins and lived with an Indian woman, Merritt loved excitement just for the hell of it. Lieutenant Arce later wrote in his memoirs that it was, at first, their intention to kill Merritt and his companions—that they were saved by the intervention of Martin and Mary Murphy who convinced Merritt to let the *Californios* keep their own saddle horses so they could return to their families. Merritt gave Arce back his sword saying, "Tell General Castro to come get his horses—if he dares."

A week later the younger Martin, on one of his trips to Sutter's Fort, was surprised to meet the commander-in-chief of the Mexican military force, General Mariano Vallejo, the most polished and educated of the *Californios*, and to learn that he had been a prisoner. Vallejo told Martin about the happenings at Sonoma, later to become known as the Bear Flag Revolt—said that a group of men had raised a flag with a bear and a star on it in the *plaza* and proclaimed California to be a republic.

On his return from Sutter's Fort the younger Martin had a serious talk with his wife. He told her about General Vallejo's imprisonment—of his concern about the Americans taking the horses from their corral—said war clouds in California were hanging low and that he feared for the safety of his father and his sister, Helen, if they were to stay on at the Ojo de Coche rancho. Mary, equally concerned, urged Martin to go to them and ask them to come stay at the Ernesto during the trouble.

On his way Martin stopped first at Mission San José where he was welcomed by the padres and invited to spend the night. There he had a chance meeting with Don Antonio Suñol, whom he had come to know at Sutter's Fort where both conducted business. Don Antonio, a distinguished early resident of the *pueblo* and sub-prefect of the district, had arrived in California in 1819 after serving as Napoleon's scribe in the Battle of Waterloo. He and his wife, Doña María Dolores Bernal, were co-owners of the nearby extensive *Rancho del Valle* with her Bernal brothers, Augustin and Juan Pablo, and her sister Doña María Pilar and her husband Don Antonio Pico. He brought Martin up-to-date on the political situation—said that he had learned the United States had already declared war against Mexico in Texas.

The next day the younger Martin arrived at the Pueblo de San José, situated beside the Guadalupe River along whose banks sycamore and willow trees grew. The village itself still consisted of a cluster of thirty or forty adobe houses and its rather primitive thatched-roof adobe church facing the plaza. Lying in the fields nearby were the bleached bones of livestock. Cattle roamed the dirt streets rutted by the heavy, wooden, oxen-pulled *carretas* that were still the *Californios'* only means of transportation other than horseback.

It was a bright, warm July day as Martin continued south on El Camino Real, crisscrossing the now dry Coyote Creek, passing hundreds of acres of grazing land marked off by park-like areas of leafy oak, stately pines and sycamores. He reached his father's *Rancho Ojo de Agua de la Coche* (usually called the Ojo de Coche) where the "old gentleman," as the family fondly referred to him, was still living in the original Hernandez adobe. Though he was unaccustomed to complaining, one of the first things he told his son was that the fleas were driving him crazy. The good news was, however, that he intended to build a new, more spacious, flea-free *hacienda* commanding a view across a little valley of the conical-shaped mountain which stands apart from the hills to the west.

This peak, which so fascinated him when they rode south with Sutter's army, served as a landmark for travelers between Monterey and the Pueblo de San José. It was called Murphy's Peak for over fifty years until the name reverted back to its original Spanish, El Toro, the bull. The elder Martin's enthusiasm for the Santa

Clara Valley and California knew no bounds. As he had said many times before, "When Father Hoecken told us about California he truly spoke well. This is an earthly paradise. I now know how Moses felt after forty years of crossing the desert to find the Promised Land. This is our Promised Land, and it only took us twenty-five years to reach it."

Although Martin was eager to get to the point of his visit, he listened patiently as his father talked about life on the *rancho*—of his pleasure with the Mexican and Indian *vaqueros* who worked the cattle—of his herd that was healthy and increasing in numbers and of his grain that grew so tall.

"Our only enemy" he said, "is the grizzly bear who sometimes kills a cow for his dinner." The younger Martin chuckled as he told his father that his *rancho's* enemy was smaller in size but greater in number—the geese who flew in from the north by the thousands and on one occasion leveled an entire wheat field.

As with his son, the elder Martin's door was always open to wayfarers who would spend a night or so on their way between Monterey and the *pueblo*. Many wrote of his warm hospitality and kindliness. The older Martin talked about his *Californio* neighbors. He said they were friendly, hospitable, deeply religious and filled with a great joy of living. On Sundays, when he rode his pony to the *pueblo* for Mass at the *adobe* church, he enjoyed their music—the choir and musicians, dressed in *ponchos* and *serapes* decorated with gold designs that hung to the floor, playing harps and violins, flutes and, of course, guitars. The women, he said, wore lace *mantillas* draped over wide combs and bright colored silk dresses.

"After my new home is completed," he said, "I want to build a chapel on the ranch and dedicate it to my patron saint."

At this point the younger Martin interrupted his father's enthusiastic discourse. "I'm afraid, my dear father, you won't be riding to the *pueblo* for a while to attend Mass." He explained to his father that California might soon be at war with the United States. The son wanted to make his point. He thought it would be safer if his father and Helen would stay with him and Mary at the ranch.

The patriarch was protesting quietly but firmly when Helen walked in and asked brightly, "Did I hear my name? I trust it wasn't in vain." She kissed her brother and her father who smiled down at her fondly.

Martin put an affectionate arm around his sister's shoulder as he told her about war being imminent—that he learned at Mission San José that the United States had declared war and there was already fighting in Texas.

He said, "Mary and I want you and father to come stay with us."

She wanted to know what her father said and when he told her, she replied with finality, "If it be the will of our father, let it be so." Then she added, "We have grand friends among the *Californios* and I know they wouldn't let anything happen to us. Besides, our father is right. He does need to stay and protect his cattle."

The elder Murphy squeezed his daughter's hand, told his son that even though he came here with a U.S. passport and had applied for Mexican citizenship, he feared no harm. He added that he truly didn't think the *Californios* would object to becoming a part of the United States because they were so far removed from Mexico City and the government there had given so little support to the province. The decision was made. The elder Martin and Helen would remain at the ranch even though most of the ranchers moved in to the *pueblo*.

CHAPTER ELEVEN

It's War Again—1846

In 1846, after Governor Micheltorena and his *cholos* had been shipped off to Mexico, there was an uneasy calm among the *Californios*. California was a plum, ripe to be plucked or, as Lieutenant Francisco Arce wrote in his memoirs, "California was like a pretty girl—everybody wanted her." The United States, France and England were casting covetous eyes in her direction. Politically, California was a house divided. There was no love lost between the new governor, Pío Pico, and General José Castro. Pico had his headquarters in Los Angeles and passed legislation giving the largest share of government funds to the south.

José Castro had other ideas. Based in Monterey, Castro controlled the customs house and the treasury. There was speculation that he planned to divide California and make Northern California a country under his rule, perhaps under a protectorate of the French or British.

The presence of Lieutenant John C. Fremont, with his sixty-two men wandering about the countryside, concerned the *rancheros*. The brash young American appeared in Monterey, where, with an introduction from U.S. Consul Thomas Oliver Larkin, he called on Manuel Castro, prefect of the district. He asked for permission to buy supplies for his party which was encamped in the San Joaquin Valley. Castro reluctantly consented. Fremont may pass through the countryside but may not enter the settlement (Pueblo de San José) and must stay away from the coast.

Less than a month later, Fremont, disregarding Castro's order, rode into the *pueblo* to meet Kit Carson and his men. In further defiance of Castro's orders, Fremont started for Monterey, camping the first night with his men and horses at the *Rancho Laguna Seca* of English sea captain William Fisher. Fisher had sailed for Baja California to move his family to his newly acquired *rancho*.

John Charles Fremont. FROM *FREMONT AND '49* BY FREDERICK S. DELLENBAUGH, 1914.

One of Fremont's loyal Delaware bodyguards. COURTESY THOMAS McENERY.

Manuel Castro was furious when he learned of Fremont's contempt for law. Castro ordered Fremont and his men to leave the country. In hostile response, Fremont led his men to Gabilan Peak above the mission village of San Juan Bautista, where he ordered them to build a log fort and to raise the American flag. His actions disturbed Consul Larkin who had hoped for a peaceful purchase of the province from Mexico.

Castro had had enough. As a show of force he brought his three hundred men to San Juan Bautista and paraded them on the plains below. Looking down from his perch, Fremont was forced to admit it was no contest. Two days later Fremont retreated, leading his force over the Pacheco Pass and up the San Joaquin Valley before continuing his march north to Oregon.

There a Marine lieutenant named Archibald Gillespie caught up with him. Gillespie said he was a confidential agent for the United States government, and had first met U.S. Consul Larkin with oral instructions from President Polk. For Fremont he carried letters from his wife, Jessie, and from her father, the expansionist senator from Missouri, Thomas Hart Benton. Gillespie gave a verbal recounting of a confidential letter to Consul Larkin from Secretary of State Buchanan and secret instructions from President Polk. From Gillespie's messages and in reading his father-in-law's letter, Fremont probably interpreted that he was to return to Sacramento, arouse the foreigners to a revolution and lead the way in making the Pacific Ocean the western boundary of the United States.

The abduction of the horses from the Ernesto and the Bear Flag Revolt were part of a rapid series of events that led up to the Mexican War in California.

In Monterey, Commodore John Drake Sloat had heard indirectly that the United States was at war with Mexico and although Sloat had as yet received no official pronouncement, after procrastinating a few days, he took action. On July 7 he sent Captain William Mervine ashore to demand surrender of the provincial government. As a twenty-one gun salute sounded out from the American vessels in port the twenty seven-star American flag was hoisted over the Customs House and a proclamation prepared by U. S. Consul Larkin and Commodore Sloat was read to the people of Monterey by Captain Mervine. Two days later Captain John Montgomery raised the flag at Yerba Buena and that same day it flew at Sonoma and at Sutter's Fort. The following week Captain Thomas Fallon raised it over the *juzgado* in the Pueblo de San José.

Captain Thomas Fallon raised the American flag over the juzgado in San José. COURTESY THOMAS McENERY.

The juzgado was an all-purpose structure, a combination of city hall, jail, and courthouse. COURTESY THOMAS McENERY.

In the midst of all this activity U.S. Consul Thomas Oliver Larkin learned about his friend General Vallejo's imprisonment and lost no time in sending a courier, John Murphy, to Sutter's Fort with a letter for him. On his way, John made a quick stop at the Ernesto to bring Martin and Mary Bolger up to date on the turn of events at Monterey.

Meanwhile, when news of the flag raising at Monterey reached the quarreling Mexican leaders Castro and Pico, they temporarily put aside past differences. With Castro's one hundred-fifty men and Pico's one hundred-twenty men, they marched southward, a day apart, to occupy Los Angeles. With Castro was a prisoner, Charles Maria Weber, whom he considered a traitor. Weber, a native of Germany, was a Mexican citizen who had declined Castro's invitation to be a captain in the Mexican Army.

In fast order John Charles Fremont appeared at Monterey with his one hundred-seventy member Bear Flag Battalion and offered his services to Commodore Robert Field Stockton who had arrived a few days earlier to replace Commodore Sloat as Commander of the Pacific Squadron. He renamed his battalion California Volunteers of the United States Army and appointed Fremont the officer in command with the title of major.

Losing no time, Stockton sailed with his men to San Pedro on the *Congress*, stopping at Santa Barbara en route to leave a detachment of men just large enough to hold and defend it. Fremont and his men followed soon after with San Diego their destination. On their arrival Fremont learned that his horses had been driven away by the *Californios*. Finding horses, which were of prime importance in wartime California, was a continuing problem for the United States forces. Fremont sent some of his men to skirmish at the surrounding ranchos and others, including Dan Murphy, on a foraging expedition across the border into Baja California. At the same time Stockton was drilling his three hundred soldiers and marines in military movements at San Pedro, Castro and Pico were drilling their men on a *mesa* above Cahuenga.

On August 6, Castro sent two commissioners to meet with Stockton. They carried a message asking for a conference and a cessation of hostilities while negotiations were pending. Stockton flatly refused, saying he would accept nothing but complete surrender. Discouraged at their inability to establish communication with Stockton, Castro and Pico departed shortly thereafter for Mexico with some of their men and the prisoner, Charles Weber. At the border they dropped Weber off to make his way back to the Pueblo of Los Angeles without food or water.

Three days later Stockton was ready to commence his march to the Pueblo de Los Angeles. A brass band led the way, followed by marines, sailors, and ox-drawn carretas mounted with guns, supplies and baggage. Just south of the *pueblo* Fremont and his soldiers, mounted on horses acquired from the *ranchos* and skirmishes in Baja California, joined them.

To the lively beat of the band, the combined forces marched jubilantly through the main street of the town. The street was lined with surprised *Angelenos* who had come out of their houses to see what was going on. It was the end of a bloodless war. The defeated had put up no opposition. Stockton proclaimed himself commander in chief and governor of the territory of California and appointed Major Fremont military commandant with the tacit understanding that he, Fremont, would become governor.

On September 2nd Stockton departed for the north, leaving Archibald Gillespie military commander of the southern area over fifty men with orders to place the village under martial law but not to enforce burdening restrictions on well-disposed citizens. Unfortunately, his instructions were not well taken. Gillespie not only treated the *Angelenos* with contempt, but his men were drunkards and behaved disgracefully. Finally, tired of being ruthlessly exploited, the *Californios*, under the leadership of General José María Flores, rebelled.

Alarmed at the seriousness of the Americans' position, Gillespie sent a courier, John Brown, better known as Juan Flaco (Lean John) to ride to Monterey to warn Stockton. He made his famous four hundred-mile ride without stopping except to change horses. When he arrived, his legs had to be pried apart in order for him to dismount the horse. On receiving news of the rebellion, Stockton lost no time setting sail with three hundred soldiers and marines for San Pedro on the *Congress*. Finding a lack of horses and fearing they were outnumbered, he ordered the skipper to continue on to San Diego.

Under orders from Stockton, Fremont (now advanced to rank of lieutenant colonel) departed by sailing ship from Monterey with his California Volunteers. At sea they met the (*Vandalia* and, in an exchange of messages, learned that there were no horses in the south. Fremont immediately issued orders to return to Monterey. Desperately in need of horses and men, he sent an urgent call to Captain Montgomery at Yerba Buena and to Ned Stern at Sutter's Fort.

A propaganda campaign to enlist volunteers was launched across the countryside, alleging that Mexican troops would arrest all Americans, take away their land, and deport them over the Sierra Nevada...pretty strong stuff considering the *Californios* had treated Americans kindly and had received them with warm hospitality in their homes. Yet, their scheme worked. Men, eager to enlist, gathered at Sonoma, Yerba Buena and, at the biggest center of activity, Sutter's Fort. The enlistees included emigrants newly arrived from the Sierra Nevada, Indians from various California tribes, Paiutes from Nevada and warriors from Chief Yellow Serpent's Walla Wallas—ready to fight with bows and arrows as well as rifles.

In the midst of the excitement the younger Martin Murphy arrived at the fort from the Ernesto to make a delivery of wheat. Martin was excited to have a reunion with the Murphy-Stephens party's good friend, Truckee, the chieftain of the Paiute nation. It was Truckee who, in the emigrants' hour of need, had directed the Murphy-Stephens party to a path across the Sierra Nevada. The chief told the younger Martin he came from Pyramid Lake with his brother, Pancho, because he wanted to help his friend, Captain Fremont.

With the urgent need for horses becoming a crisis, Fremont solicited Charles Weber for help. Weber, who had already raised volunteers for him, responded with dispatch, scouring the countryside, raiding the *ranchos*, and combing the pastures and corrals.

He was assisted by a group of men including young John Murphy. Although Weber was considered ruthless, unknown to his critics he had made a meticulous list for his commanding officer designating the number of horses and saddles taken from each rancho.

Weber, with three hundred horses cached at his *Rancho Las Animas* below the *pueblo*, was on his way to deliver them to Fremont at Monterey. In this endeavor he was joined by journalist Edwin Bryant. The two camped at Martin Murphy's *Rancho Ojo de Agua de la Coche*—a convenient arrangement for Weber as it gave him an opportunity to see Helen Murphy whom he was courting and would eventually marry.

When Manuel Castro, commandant of the Northern California forces, got wind that two bands of Americans were moving three hundred horses to deliver to Fremont at Monterey, he was determined to intercept as many as he could. Captain Charles Burass (or Burroughs) of California Battalion D came from Sutter's Fort with twenty-two men including ten Walla Walla and two Delaware Indians, two Paiutes (Truckee and his brother Pancho) and three hundred horses. He

arrived in San Juan Bautista the same day, November 15, 1846, as Capt. Bluford "Hell Roaring" Thompson with his California Battalion G. Fremont meanwhile had already proceeded to Monterey.

Among Thompson's company was nineteen-year-old Dan Murphy who had first served in Company D, California Battalion and was now orderly sergeant and interpreter for quartermaster companies C, G, and H. Dan was elated when he discovered that his old friend, Joseph Foster, member of the Murphy-Stephens party, was now a captain with Company C. The two upstanding young men were misfit members of this motley crew of thirty-five men including runaway sailors still in their sailor suits, ranchers in buckskins, English and Germans—all escaping from somebody or something.

Information brought by couriers did not bode well for the Americans. Manuel Castro, whose army outnumbered them two to one, was waiting to ambush them where the road crossed the Salinas River. On top of that they learned that U. S. Consul Thomas Oliver Larkin had just been taken prisoner at Don Joaquin Gomez's neighboring *Rancho Vergeles*. Pressured by Thompson, Burass, took the lead, mounted on Fremont's gray charger, Sacramento.

On the border of the *Ranchos Natividad* and *Vergeles* he met Castro head-on for what would become known as the Battle of *Natividad*. During the brief encounter, the Walla Wallas on their Cayuse ponies taunted the *Californios* and returned their fire with a shower of arrows. Afterwards the Delawares, using their tomahawks, scalped the four *Californio* victims for trophies.

Captain Burass was shot in the chest and buried on the hill at the Los Vergeles rancho along with three other Americans. Dan Murphy escaped without injury, but was deeply saddened to learn that while reconnoitering, Joseph Foster had been killed. An inscription carved on the tree where he fell and under which he was buried read: Foster—1846.

Tom Hill, a Delaware, and Charles McIntosh, a part Cherokee or Delaware, acting as couriers, sped to Monterey to inform Fremont of the battle. Fremont returned and moved the entire battalion to San Juan Bautista where he trained the men on the streets of the mission village. Early morning on November 30, during a heavy downpour, the California Battalion departed the mission grounds with five hundred horses salvaged in the *Battle of Natividad*. Their only pretense of a uniform was a navy blue shirt with a white star on each corner of the collar.

Meanwhile, in San Diego, Commodore Stockton received a communiqué from General Stephen Kearny reporting that he was in California, eighty miles distant with his Army of the West, and needed help. Stockton responded immediately by sending an escort of fifty men under the command of Captain Archibald Gillespie. As the combined units continued on their way, at San Pascual a bloody battle ensued—Stockton and Gillespie were badly injured and transported to San Diego on stretchers.

General Stephen Watts Kearney. COURTESY AN AMSCO SCHOOL PUBLICATION.

By January 8, 1847, the two officers had recovered. Stockton ordered the march to Los Angeles to commence. They met the enemy on the banks of the San Gabriel River, succeeded in crossing it and driving the enemy back. After fighting the next day on the *mesa*, General Flores gave up, turned the command over to Andres Pico and left for Mexico. Three *Californios*, including Juan Bautista Alvarado, approached the American camp bearing a flag of truce, surrendered the Pueblo of Los Angeles and agreed to respect property and persons. On January 10 the Americans raised their flag over the Pueblo de Los Angeles.

While all this was happening Fremont and his battalion were still sloshing through the mud and swollen streams, over ground so saturated with water that many of their horses gave out from exhaustion. As Dan Murphy would later describe the scene to his father, the elder Martin, "We were a sad looking group—muddy and wet from having traveled continuously through rain and cold since we left San Juan Bautista." The battalion finally reached Santa Barbara on December 27.

Without opposition, they occupied the town and raised the United States flag. During their week's stay Fremont received several communiqués from Kearny in which he made clear who was in charge—that he was general of the Army of the West. Fremont, nevertheless, proceeded with his plans to negotiate a peace with the *Californios*. On January 11 a few miles north of San Fernando, he received a message from Stockton informing him of the Battle of *La Mesa* and the capture of the Pueblo of Los Angeles. The California Volunteers camped that night in the San Fernando Mission buildings.

After the long, slow journey, suddenly there was action. José Jesus (Tortoi) Pico, who vowed to repay his debt to Fremont for his compassion in commuting his death sentence, set out to find the *Californios* and open negotiations with the leaders. He returned to Fremont's camp bringing two officers of Andres Pico's army for preliminary peace talks.

The next day Fremont, joined by several of his men, rode up to an empty *adobe* house at the north end of the Cahuenga Pass where he met with Andres Pico. Fremont drew up the terms of the Capitulation of Cahuenga between the United States and Mexico, and it was signed January13, 1847. The treaty stated that the conquered could return to their homes if they promised not to take up arms again and promised to conform to the laws of the United States, and that they would have the same rights as those of all United States' citizens. No prisoners were taken nor were the *Californios* required to take an oath of allegiance to the United States.

When they received a copy of the treaty, Stockton and Kearny were furious at Fremont's leniency. Nevertheless, when Fremont and his battalion marched triumphantly through the Cahuenga Pass to the Pueblo of Los Angeles, it was a feeling his men would never forget and one Dan Murphy liked to reminisce about in later years.

While all the action was taking place in the south there was concern among the naval officers that those Americans left in San José might be maligned by the

Californios. Captain Hull commissioned Charles Weber captain and John Murphy lieutenant to form a volunteer company, the San Jose Militia, for their protection in case of trouble. The only trouble that developed was the Battle of Santa Clara, or as it was later nicknamed, the Battle of the Mustard Stalks, because the tall native mustard was in bloom.

Although the *Californios* in the north were not opposed to becoming part of the United States—so removed from Mexico they had no close ties with the government—they objected to the high-handed actions of the Americans.

Don Antonio Suñol, prefect of the district, had written to naval Captain John Montgomery at Yerba Buena on behalf of the *rancheros* protesting the ruthless rustling of horses, a practice which left them without stock to work their *ranchos*. Montgomery misread Don Antonio's intentions and put the blame on the *Californios*. That was enough for Don Francisco Sanchez of Rancho Buri Buri, who was already angry because of the Americans' invasion of his brother's home. He assembled a group of *rancheros* and captured Lieutenant Bartlett, *alcalde* of Yerba Buena, while he was conducting a cattle-raiding party near Mission Dolores.

Marines of the Yerba Buena post led by Captain Marston joined Captain Weber and Lieutenant John Murphy with the San Jose Mounted Volunteers at Mission Dolores. They then proceeded down El Camino Real—the Marines on foot. When a call for volunteers went out, Lieutenant John Murphy and five of the San Jose company's privates were among the first to sign up.

The two forces, evenly matched in manpower, met on the oak-covered plain within sight of Mission Santa Clara. The *rancheros* had the advantage of expert horsemanship and fine mounts, giving them maneuverability. The Americans, aside from their cannon carriage becoming stuck in the mud, turned the *Californios* back with musket and cannon fire, bringing a quick ending to the so-called battle. There were no casualties and within two hours from the time the Americans spotted the *Californios*, it was all over.

Captain Marston, a Captain Smith of the Yerba Buena Volunteers, and British Consul James Alexander Forbes, who was invited to serve as interpreter, met the next day on the plain with Don Francisco Sanchez and his U.S. captives, Lieutenant Bartlett and two of his men. Sanchez told the American contingent that the *rancheros* were not in arms against the American flag. They were not carrying a flag and they were not associated with any Mexican group. The Americans agreed to repay them for their losses, with the assurance that their property was secured to them in the future.

On February 2, 1848, the Mexican War officially ended with the signing of the Treaty of Guadalupe Hidalgo at Chapultec Castle in Mexico City.

Manuel de la Peña y Peña Presidente
interino de los Estados Unidos Mexicanos

A todos los que las presentes vieren SABED:

Que en la Ciudad de Guadalupe Hidalgo se concluyó y firmó el día dos de Febrero del presente año, un tratado de paz, amistad, límites y arreglo definitivo entre los Estados Unidos Mexicanos y los Estados Unidos de América por medio de Plenipotenciarios de ambos Gobiernos autorizados debida y respectivamente para este efecto, cuyo tratado y su artículo adicional son en la forma y tenor siguiente:

En el nombre de Dios Todopoderoso:	In the name of Almighty God:
Los Estados-Unidos Mexicanos y los Estados-Unidos de América, animados de un sincero deseo de poner término á las calamidades de la guerra que desgraciadamente existe entre ambas repúblicas, y de establecer sobre bases sólidas relaciones de paz y buena amistad, que procuren recíprocas ventajas á los ciudadanos de uno y otro país, y afiancen la concordia, armonía	The United-States of America and the United Mexican States, animated by a sincere desire to put an end to the calamities of the war which unhappily exists between the two republics, and to establish upon a solid basis relations of peace and friendship, which shall confer reciprocal benefits upon the citizens of both, and assure the concord, harmony, and mutual confidence

The Treaty of Guadalupe y Hidalgo signed at Chapultepec Castle near Mexico City. One of the stipulations was that the Californios should retain their land. Unfortunately Congress passed an act stipulating that the landowners would have to produce proof of ownership making it necessary to hire lawyers. FROM EL TRATADO DE GUADALUPE HIDALGO 1848.

CHAPTER TWELVE

The Gold Rush

With the arrival of the new year, 1848, all was running smoothly at the Ernesto. A new baby named Mary Ann had been added to Martin and Mary's growing family, the crops were productive and the cattle had plenty of feed. Martin made frequent trips to Sutter's Fort with wagonloads of wheat and traded with Sutter in horses and oxen. But on a trip he made toward the end of January he received news from the Swiss captain that would effect the rest of his life, as well as the lives of all the Murphys, and would change the course of California's history—it was the discovery of gold at the sawmill John Marshall was building for Sutter on the American River in the Coloma valley.

Sutter's Mill.

Sutter's Mill, Coloma. COURTESY CALIFORNIA HISTORY SECTION, CALIFORNIA STATE LIBRARY.

This event, which would prove so earth shaking, happened by accident. While in the course of work on the sawmill Marshall found it necessary to deepen the millrace in order to increase the flow of water necessary to turn the wheel. In so doing, he loosened a large bed of mud and gravel, washing it down to the race.

At that point, as Marshall was later to recall, "My eye was caught by something shining in the bottom of the ditch. I reached my hand down and picked it up; it made my heart thump, for I was certain it was gold. The piece was about half the size and shape of a pea. Then I saw another and another." Subsequently, while he tested the metal to be sure of what he had found, work continued and the force of the water running through continued to loosen the rocks, gravel and sand, revealing more gold.

James Wilson Marshall, the man who started it all when he picked up the gold nuggets from the millrace at Sutter's mill. Ironically, both he and Sutter ended up poor. COURTESY CALIFORNIA HISTORY SECTION, CALIFORNIA STATE LIBRARY.

Martin returned home and called excitedly to Mary. She came in a hurry, wondering what her normally quiet, steady husband could want. Breathless, he told her the news that gold was discovered at Captain Sutter's sawmill. Mary looked at him with a puzzled expression, wondering aloud, "Would that mean that there would be more gold there, Martin?"

Restraining a smile, he said, "Yes, Mary, and maybe a lot of it. But, mind you, mum's the word. The Captain and John Marshall want to keep it a secret."

That kind of a secret was hard to keep; still the news spread slowly at first. Even when the charismatic Mormon leader, Sam Brannan, printed an account of the discovery in his new *California Star* newspaper, it created little interest. It was a different story, however, when he returned from a trip to Coloma in early May and ran down Montgomery Street, waving a quinine bottle filled with gold dust in one hand and his hat in the other shouting, "Gold! Gold! Gold from the American River." Brannan's publicity stunt, which was aimed at business for the store he opened while there and another he was planning to open at Sutter's Fort, ignited a spark that before long would burn out of control and bring great wealth to Brannan.

When Colonel Richard Mason, commanding officer of the First Dragoons and military governor of California, first heard the extravagant accounts of gold findings he was a doubting Thomas. Finally, in July of 1848 he made a trip to the mines with his chief of staff, Captain William Tecumseh Sherman. They were astounded at what they saw. He immediately wrote a report to the Secretary of War explaining he had not written sooner because he had dismissed as tall tales the stories about gold found on the American River until he actually saw it.

"I struck this stream at the wastings of Suñol and Company," he said. "They had about thirty Indians employed whom they pay in merchandise." He estimated there were approximately four thousand miners, that half were Indians and that they were taking in between $30,000 and $50,000 per day. He said that no capital was required to obtain this gold, adding, "the laboring man wants nothing but his pick and shovel and tin pan with which to dig and wash the gravel, and that many frequently pick gold out of the crevices of rocks with their butcher knives in pieces from one to six ounces."

A month earlier U.S. Consul Oliver Larkin wrote to Secretary of State James Buchanan, "I have to report to the State Department one of the most astonishing excitements and state of affairs now existing in this country that perhaps has ever been brought to the notice of the government. On the American fork of the Sacramento and Feather rivers there has been within the present year discovered a vast tract of land containing gold in small particles." His letter and Colonel Mason's reached Washington by November in time for President James Polk to incorporate the news in his message to Congress on December 5. The President's reference to the gold discovery, combined with the wide publication of Larkin's letter and those of others, provided the impetus for the really big push.

The effect of the news snowballed until it seemed as though the whole world had become infected with gold fever. Over two hundred thousand gold seekers would make their way to California. From the eastern United States and the Middle West the Argonauts, as they were called, crossed the plains by the thousands in their covered wagons. Others took the shorter trip by ship to Chagres and made the dangerous crossing of the Isthmus by canoe and mule train to Panama City where, during the wait for another ship to take them to San Francisco, cholera took its toll. They arrived at Yerba Buena, recently renamed San Francisco, to discover the mass exodus to the Mother Lode had left the once growing town half empty. Partially built houses stood out like skeletons.

In San Jose the mayor, who was disenchanted with his job, took off for the diggings, leaving his puzzled sheriff in charge with ten incarcerated Indians, some of whom had been charged with murder. With a shrug of the shoulder, the sheriff closed the *juzgado* and, prisoners in tow, headed for the hills. This proved to be a smart move because the Indians mined enough gold to make him a wealthy man.

Don Luís María Peralta, San Jose's most respected citizen, counseled his sons to go to the family's forty-eight thousand acre *Rancho San Antonio* and raise grain, that it would be their best gold field. The elder and younger Martin Murphys subscribed to this theory and profited in raising grain and cattle. A devout, religious man, Don Luís said, "God gave the gold to the Americans. If he had wanted the Spaniards to have it he would have let them discover it before now." At the same time there were Protestant preachers proclaiming that God had waited to reveal the gold to the Americans until after the "Popery," as they called the Catholics, was no longer in control.

The discovery of gold proved to be the turning point in the fortunes of Charles Weber as well as of his future

brothers-in-law, John and Dan Murphy. Weber first formed the Stockton Mining Company in which John and Dan were among several stockholders. Getting a jump on the masses of prospectors who were to come in the spring of '48, they started at Chowchilla Creek near Hangtown (Placerville). Eventually they moved down to the Mokelumne River where they hit a rich strike in the area later called the southern mines. The area's location south of the Cosumnes, closer to his settlement, Tuleburg, assured its position as the starting point for prospectors whose numbers were increasing at an astonishing rate.

At this point Weber decided to disband the company. He told the Murphys he had concluded the real gold was to be made in supplying the miners with food, mining equipment and medical supplies. John had run a camp selling merchandise for Weber and, recognizing the wisdom of Weber's decision, the brothers decided to open their own merchandise mart. They made an arrangement with Weber who would provide them with tools, cooking utensils and clothing when they found a location. They hoped to buy flour and meat from their brother, Martin.

Weber arrived back at Tuleburg (Stockton) to find it looking much the same as when he left. There was little more than Joe Buzzell's tavern in a log cabin, tule-roofed huts and tents. This, as it turned out, was the calm before the storm. Soon business at Weber's store was rushing and highly profitable with prices as high as two hundred dollars for mining tools such as a pick, pan and shovel, butcher knives for from ten to twenty five dollars and cradle washing machines fetching from two hundred to eight hundred dollars. Woolen shirts were worth fifty dollars each and boots and shoes cost from twenty-five to one hundred-fifty a pair. Crackers were a dollar per pound, pickles, sixteen dollars per quart, canned meats and sardines were priced at sixteen dollars a can.

Through the Golden Gate at San Francisco, sailing ships flying the flags of France, the Sandwich Islands, the United States, and Chile, swarmed the bay. Many would lie deserted after captains and crews, bitten by the gold bug, jumped ship to join the prospectors. By the end of 1849 some seven hundred ships brought over forty thousand gold seekers to San Francisco Bay.

To reach the main departure points for the mines located at Sutter's Fort and Weber's Embarcadero, the prospectors had a choice of traveling by ship up the Delta, or by foot over the Altamont Pass with packs and tools strapped to their backs. If he could afford it, the prospector might buy a mule to carry his baggage. There were few wagon roads so he usually joined a pack train consisting of seventy-five to one hundred animals led by a white stallion or mare wearing a bell as they crossed the Livermore and Altamont passes.

The alternative was to board one of the poorly-equipped sailing ships and to be packed in like a sardine. The latter was not a pretty prospect. It often took fifteen days for the ship to snake its way through the maze of marshy channels—its passengers tormented by the mosquitoes living in the tall tules where, many times, they would have to anchor for the night.

The discovery of gold proved to be a boon to Captain Weber, not only in acquiring great wealth in ore and in merchandising, but also, with the influx of miners passing through to the gold country, it was attracting settlers for his new city. At first, even though he had offered free building sites along with a parcel of agricultural land, most were reluctant to settle there because of their fear of the hostile Indians in the area. Taking a page from the book of Captain Sutter, Weber made friends with the Indians' chief, the feared José Jesus. By making generous gifts he succeeded in keeping the Indians appeased.

James Carson, one of the more reputable members of Stevenson's Regiment of New York which had come for the Mexican War and stayed on to reap a golden harvest, wrote a colorful description of the Embarcadero:

"The tall masts of barques, brigs and schooners" he said, "were seen high-pointed in the blue vault above while the merry 'yoho' of the sailor could be heard as box, bale and barrel were landed on the banks of the slough."

Another writer of the same regiment, Edward Gould Buffum, wrote, "The town of Stockton is the great mart through which flows the whole transportation and travel to the placers of the Stanislaus, Mokelumne, Mariposa, Tuolumne and King Rivers and the various dry diggings lying between them. The town was located upon a slough or rather a succession of sloughs which contained back waters formed by the junction of the Sacramento and San Joaquin rivers." With the rapid turn of events caused by the gold rush, Weber recognized the need for better transportation between his embarcadero and San Francisco. He purchased a two-masted sailing ship, The *Maria*, to trans-

port supplies, and the following year the first mail packet ship, the *Emil*. After 1850, steamers took over.

Meanwhile, as the gold frenzy grew, so grew activity at the younger Martin's rancho on the Cosumnes. With the demand for wheat as well as beef to feed the miners and with the responsibility of his growing family which now numbered seven children, Martin chose not to gamble on the gold fields. He was showing an increasingly handsome profit on his deliveries of grain and cattle to New Helvetia, yet he was appalled to see what was happening to his friend, Captain Sutter. Sutter's Mormon and Indian help, who stayed with him at first, were seized with the burning fever that "raged in the souls of men" and left to mine for themselves. His own efforts at acquiring gold were a failure. He tried hiring Indians to dig for him, but traveling saloons followed them and enticed the Indians to squander their take on liquor and gambling. It was beyond his control, because he was one of the saloon's best customers.

At the same time his fort was in a chaotic state. As he later wrote, "Stealing began. Land, cattle, horses, everything began to disappear." They trampled his crops and grain fields so his horses could not graze. His salmon barrels vanished along with the fort's bells, cannon, and gate-weights. Whisky bottles were strewn about in heaps and, in addition to the drinking, gambling and whoring, squatters moved on to his land. As Sutter's partner in the sawmill, James Marshall, referring to the squatters, said, "They left honor and honesty at home." The only sign of order was the corps of men under Corporal Edmund Bray, a member of the Murphy-Stephens party, who arrested soldiers-turned-deserters. The army and navy had given up on trying to apprehend the men who had gone AWOL. Inevitably, the men assigned to arrest them would desert to the mines themselves.

Sutter ran into obstacles in acquiring title to what was once called his "citadel." The dream of an agricultural empire on his fifty thousand acres was going down the drain. Even John Fremont, whom he had befriended, turned traitor. It may have been California's most exciting period, but for Captain John Sutter, it was its most degrading. He would eventually lose his land when the Supreme Court ruled against him and gave the land to American squatters. He was reduced to living on a modest government pension.

The life of a miner was not an easy one. The work was strenuous, often requiring miners to dig in riverbeds with water up to their hips. Their living conditions in crude huts or tents were deplorable and they suffered extreme heat in summer, freezing cold in winter. Despite the great fortunes of some, gold was not within every man's reach. Not even James Marshall who started it all when he found those first gold nuggets on the American River, succeeded as a miner. Furthermore, most miners came not as settlers but to make a quick kill in the mines. Law and order were improvised.

In 1848 there was no official form of government. No laws had yet been passed to control or tax the miners because much of the land was in public domain. Each camp worked out its own code of laws and elected an *alcalde*, but no man's life was safe. In time the rivers, creeks and gulches were all claimed and as the rivers became overworked, men had to labor twice as hard to find the gold. After working like a dog, a man would go into Sonora with his poke of gold and most times blow it all on booze, gambling and women. There were also vultures who would get a man drunk, rob him or cheat him at gambling, or hit him over the head and rifle his bags.

The biggest problems occurred in the southern mines. The Americans resented some of the other national groups whose largest numbers were made up of Mexicans coming up from Sonora, following the Anza trail of 1776 and of Chileans who came by sea to San Francisco. They were treated abominably by the Americans—often victims of violence. Anyone who spoke Spanish was considered despicable. The problem, in part, was due to jealously because of the superior mining skills the Sonorans and Chileans possessed. They would sink holes and when they struck the gold stratum would follow the veins underground. The Americans tried to have the Sonorans banished. But Military Governor Bennet Riley, in their defense, said that the land was in public domain and that all were trespassers. Equally scorned were the Chinese. Even though they were American, the Southerners, who brought their slaves and claimed stakes in their names so that they controlled large territories, were resented.

Only a few of the *Californios* were lured by the glitter of gold, at least at first. Skeptical about the Americans, most of whom had treated them badly, they preferred the peace of their homes to the promise of the gold fields. Among the exceptions were Don Antonio Suñol, his brother-in-law, José María Amador and his son-in-law, Pierre (also called Pedro) Sainsevain (for whom Pedro's Bar was named). After mining successfully for two weeks, they returned to San José with their fortunes. Antonio Coronel was another success-

ful *Californio* who came up from Los Angeles with a party of about thirty men. At the San Joaquin River they met a Franciscan *padre*, José María Soares del Real from Mission Santa Clara, who had great success mining at the Stanislaus. The jovial *padre* was on the verge of being banished to Mexico for unpriest-like conduct. Still his advice served Coronel well when he and his party proceeded to the Stanislaus.

Back on the Cosumnes, Martin, Jr. was doing a thriving business selling cattle and wheat to the miners. The steady stream of travelers stopping at the Murphys' consisted mostly of miners who would readily pay four dollars for a night's lodging and a meal. It was a welcome change when Bayard Taylor, a correspondent for the New York Tribune owned by Horace Greeley, arrived. He later wrote in his book, *El Dorado or, Adventures in the Path of Empire*, of his visit to the Ernesto and of being met by an Indian youth who tied his horse to a haystack after which he was welcomed by a friendly Mrs. Murphy. "You must be hungry," she said and set about to make some biscuits and roasted a piece of beef on a wooden spit.

That night he shared a room with a company of gold diggers on their way from the Yuba to winter on the Mariposa, the ranch of John Fremont. The author found the miners at one end of the house where they lay rolled in blankets—their forms barely discernible through the smoke sent up by the rain-soaked wood on which their fire was made. He talked with them about the prospects of mining on the different rivers.

The next morning Taylor discovered that his host was the son of Martin Murphy whose hospitality he had enjoyed at his hacienda on Monterey Road in south Santa Clara Valley. In his account of his stop at the Ernesto he said that he was impressed with the horsemanship of the Murphys' sons, James, Martin, and even eleven-year-old Patrick and nine-year-old Barney, who could ride and throw a lariat as well as any *Californio*. This, he said, was high praise, because the *Californios* were considered the best horsemen in the world.

A story typical of the times was related during a visit from Martin's brother, Jim. He had had a thriving lumber business in Corte Madera providing the timber for Leidesdorff Wharf at Yerba Buena (recently renamed San Francisco) and for the growing number of homes being built there. Unfortunately, he said, he could no longer get workmen. They had all deserted for the gold fields. He decided to have a go at mining himself but returned in a few months saying prospecting was not for him. With a family to think about, he was going to see their father about buying land near him and raising cattle. The Murphy family ties were strong.

It wasn't long before Martin's younger brothers, John and Dan, came in, heavily bearded, dressed in the familiar prospector garb of red flannel shirt, heavy trousers tucked into their high boots, wide leather belt and slouch hat. Martin and Mary were as anxious to hear about their experiences as the brothers were bursting to tell. John, the more outgoing of the two, said, "We were lucky to get an early start with Captain Weber's group." He talked of their successes at Hangtown and on the Mokelumne River, saying that at that point Captain Weber determined he would turn back and go into merchandising at Stockton.

Dan, although only twenty-one-years old, was the businessman. "We decided" he said, "that the big money is in selling merchandise to the miners. For now we are joining the rest of our group on the Cosumnes and will follow the creeks and rivers until we find the right place to open a trading post. Captain Weber is going to supply us with tools, cooking utensils and clothing." Martin concurred completely with his brothers' plans. They then asked if Martin would provide them with flour and beef.

"That's a foolish question" he said, "Of course, I will."

A plaque believed to be the location where the Dan and John Murphy had their trading post and reaped a golden harvest. COURTESY MARJORIE PIERCE.

As they were about to depart, John remembered that he had had a bit of news about John Sullivan, their old friend from Missouri who crossed the plains with them. "We learned from a prospector that John discovered a rich mine in a canyon off the Stanislaus River—called Sullivan's Diggings— that he took out $26,000 in gold and that he planned to invest in San Francisco lots."

After fond *abrazos* all around the brothers were on the road again with their Stockton mining group— working the Cosumnes, Dry Creek, Sutter's Creek and the Mokelumne, Calaveras and San Antone Rivers, finding gold in most of them. One of their members, Henry Angel, decided to part company when they reached a small stream that interested him. He opened a store around which a small town grew up that became known as Angel's Camp. Another member, James Carson, with several of the group, moved on to another site which yielded an impressive profit. It became known as Carson's Creek. Dan and John continued eastward over a hill to Coyote Creek, a tributary of the Stanislaus River, where they found rich gravel. Here they established the first Murphy camp known as the Old Diggings, a name the Mexicans found difficult to pronounce and changed a few years later to *Vallecitos*, meaning, Little Valleys.

Although they were already doing well, an Indian gave the Murphys an invaluable tip directing them up the Coyote north to a pine covered valley with a stream meandering through that had first been prospected by a man named Staudenberg. They had been acquiring gold along the way, but this turned out to be the big one. With supplies provided by Charles Weber and their brother, Martin, they set up a trading post that was first called Murphys New Diggins, then Murphys Rich Diggins, Murphys Flat, Murphys Camp and, finally, it became simply Murphys, the name of the little town that grew up around it.

The Murphy men dealt easily with many of the Indians and made arrangements to exchange merchandise for gold with members of a *rancheria* nearby. Still, it was John's marriage to the lovely Pokela, sister of John's friend, José Jesus, whom he first met at Knight's Landing on the Stanislaus River, that made a difference. Because he had an Indian wife, John was referred to by some as a "squaw man."

At the Murphy camp the gold soon poured into the store—as much as twenty-five pounds a day. Pokela was a loyal wife and, with her family ties, was of great support to John, keeping him informed as to what was going on with the Indians. Pokela's big concern was to protect her husband. On one occasion when John, accompanied by Pokela, went up to spend a night at his post near some unfamiliar Indian tribes, attempts were made on his life. In letters written to the *Boston Daily Evening Transcript* in 1849, W. E. Warren told of

Pokela, who, realizing the danger to her unsuspecting husband, dispatched a trusted Indian on horseback to his camp with a request for arms, ammunition and men. Fortunately, the anticipated outbreak was quelled and the tribe fled to the mountains.

By the end of 1848 Dan told his brother he had decided to return with his booty to their father's home in the Santa Clara Valley. Like most of the Murphys he was land hungry. He now had the wherewithal to satisfy that appetite, which he would do in a big way. By the time of his death in 1882 he was reputed to own more land and more cattle than anyone in the country.

The word of the Murphy's success spread, attracting gold seekers in large numbers. The town became, as one visitor described it, a little metropolis of tents and temporary buildings. A broad street ran down the middle lined with larger tents offering mining supplies, food and other necessities.

In August of 1849 Rev. Walter Colton, U.S. Navy chaplain, Monterey's first *alcalde*, and co-founder of the state's first newspaper, *the Californian*, visited Murphys while traveling through the Mother Lode on a quest for gold.

He later wrote in his book, *Three Years in California*, of John Murphy inviting him and his companions into his tent. "Although pitched in the midst of a tribe of wild Indians," he said, "he set before us refreshments that would have graced a scene less wild than this." Murphy told him that he knocked down two bullocks a day to furnish his Indian workers with meat. When Colton commented that he had seen no intoxicated Indians, John said, "I allow no liquor in the camp. Only a few days ago a trader brought in a barrel of rum.

I gave him five minutes to decide whether he would leave the grounds or have the head of the barrel knocked in."

"Kinship ties with the tribe" he said "were strong." Pokela's brother, the chief José Jesus, had a strong influence with the various groups and assumed their leadership. He had been among the Miwok Indians moved in the 1830s to Mission Santa Clara where he received an education.

In 1870 Frederick Hall wrote in his book, *A History of San Jose and Surroundings*, "The remarkable fortune of a person in mining, now a resident of San Jose, was so much like the Arabian Night's tales, that I cannot refrain from detailing it." He said John Murphy "had, at one time, nearly one hundred and fifty [Indians] working for him" and that "his influence with them was as productive of gold as the exclamation of "Open Sesame" by Ali Baba, at the cavern door of the forty thieves… He had buried in the ground nearly two million dollars worth of gold. But at that time, it was sold cheaply. Coin was scarce and most of the people, for a long while, did not know its real value. It was customary to sell the gold at four dollars per ounce, which was intrinsically worth over sixteen."

By December John knew it was time for him to leave the gold country. It was a decision he had been agonizing over for some time. The parting scene was made more poignant when Pokela, without histrionics, said she understood. John's heart was as heavy as the seventeen sacks of gold on the six mules he was leading back to civilization, but the book on this period of his life was now closed.

CHAPTER THIRTEEN

San Jose—California's First Capital

In spite of the younger Martin's financial success at the Ernesto, he and Mary Bolger were concerned about the turn their lives had taken since the advent of the Gold Rush. First of all, they feared the influence on their children of the miners who stopped at the Ernesto overnight on their way to the gold fields. Not even a charge of $4 per night for a bed, meal and care of a horse, would deter them from coming. Also, the damp weather along the river was a haunting memory for them of the malaria crisis on the Missouri River in which they lost their nine-month-old daughter.

In late spring of 1849, Martin made a trip to the Santa Clara Valley that was to change the course of their lives. He was planning to buy a herd of cattle for delivery to the miners at the gold fields, but when the seller refused to honor the agreed-upon price the deal fell through. While there, always on the lookout for land, Martin happened to see the *Rancho Pastoria de las Borregas* (Sheep Pasture), about eight miles up El Camino Real from San Jose. When he learned that its owner, Mariano Castro, wanted to sell half of it he took his father, whose opinion he valued, to see it.

The patriarch weighed it with a critical eye. After a long pause, he said, "From a selfish point of view, it would make me happy to have you, Mary, and the children close by. Still, in truth, I do believe it is a fine piece of property."

That was all the younger Martin needed. With the gold he carried to buy the cattle, he decided to purchase half the *Borrega* covering over four thousand acres, for $5 per acre. They rode in to the *pueblo* to see an attorney named William Wallace, soon-to-be son-in-law of California's first governor, Peter Burnett. As Wallace drew up the deed of sale, he looked at Murphy, shook his head and said, "Young man, you have one hell of a lot of nerve to settle in that wild country."

Peter H. Burnett elected first governor of California. COURTESY CALIFORNIA HISTORY SECTION, CALIFORNIA STATE LIBRARY.

On his return to the Ernesto an apprehensive younger Martin, who always consulted his wife first on business negotiations, gently sat Mary Bolger down. "I have something to tell you." Noting her concerned expression he hurried to tell her about the *Borregas* purchase.

He needn't have worried. Relieved that the news was good, Mary Bolger's face lighted up with a broad smile and her voice rose as she said excitedly, "I think that's wonderful. The children have had so many colds; the climate there should be healthier for them. "Besides," she added, "we won't have to bother with the miners who are starting to overrun the ranch, stealing our horses and our crops."

Martin, who shared his wife's devotion to the Catholic faith, added, "Mary, for the first time we will be able to take the children to Mass in a church. We won't be far from Mission Santa Clara. Maybe they'll even be able to attend a school." Respectful of his wife's canny sense of property values, he promised, "My dear, this is the last piece of land I will buy without talking to you first."

Soon ready to start off on another adventure, the children were exuberant as they packed their belongings and helped load the wagons. The four boys now had three younger sisters, Lizzie Yuba, born at the camp on the Yuba River, and Mary Ann and Nellie born at the Ernesto. Martin had made arrangements for the family to live in the Hernandez *adobe* on the outskirts of San José until the makings of their new home arrived and were made ready.

It was a new experience for the children to have neighbors and their first opportunity to make friends with the *Californios* and to learn their language. The boys, having spent their early years in Quebec where French was the first language, quickly adapted to Spanish.

They were especially taken with their closest neighbor, Don Luís María Peralta, a legendary figure who had lived under the flags of Spain and Mexico, the Bear Flag and, now the Stars and Stripes of the United States. He told them tales of his coming as a boy in 1776 with the Anza party, of walking all the way from Tubac, Mexico to Yerba Buena (as San Francisco was then called), of experiencing the bitter cold while crossing snow-covered mountains and the searing heat of the desert.

Spreading out around Peralta's comfortable home were acres of orchards where the younger children could run and, when the pears and peaches ripened,

they could eat as many as they pleased. Their biggest excitement, however, occurred when the authorities came to the house in search of Mariano Hernandez, who was accused of murdering an American named John Foster, brother of Joseph Foster, who had been a member of the Murphy-Stephens Party.

At the Constitutional Convention in Monterey, the Pueblo de San José de Guadalupe, (shortened to San Jose in 1847) was chosen to be the first capital of the new state due to the determined efforts of James Frazier Reed of the Donner Party and Charles White, a politically active young man from the Middle West. The meeting of the first legislature was already in session.

Although their plans for a proposed capitol had gone awry, White and Reed succeeded in raising thirty-four thousand dollars from nineteen supporters to purchase a hotel that was being built on the plaza by Pierre Sainsevain, a prominent *Californio* delegate to the convention. In the meantime, meetings were held at Isaac "Uncle Ike" Branham's *adobe* at the south end of the *plaza*.

James Frazier Reed. He fought tirelessly with San Jose's mayor, Charles White, to make San Jose the first capital of California. COURTESY SCOTTWALL ASSOCIATES.

Unfortunately torrential rains hit this part of California that winter making it necessary for the first delegates, coming by steamboat to Alviso, to travel by stage to the *pueblo* over a six mile area covered by water. Fewer than half the delegates were able to arrive on time for the opening session on December 15, 1849. When the stagecoach from San Francisco was forced to discontinue its run because of the inclement weather, the governor-elect, Peter H. Burnett, made the trip in a primitive mule-drawn *carreta*—not the most comfortable of conveyances.

Those who did succeed in making it to the meeting on time found housing at a premium. The best place to stay was the boarding house of Anna Marie "Grandma" Bascom, which her son had named Flapjack Hall because of the popularity of her pancakes. John Charles Fremont and his wife, Jessie, among other dignitaries, were paying guests, as was the younger Martin on one of his cattle-buying trips to the *pueblo*.

The legislators, described as a lively, hard-drinking crowd, acquired the epithet, "The Legislature of a Thousand Drinks," from the speaker's habit of saying at the end of a long day's session, "Let's go have a drink, boys, let's have a thousand drinks." Their propensity for alcohol, of course, was an exaggeration. They were, in the main, a dedicated group and their accomplishments were notable.

After the quiet life in the wilds of the Sacramento Valley, the Murphy children experienced a strange new world. While the little *pueblo* of *adobe* houses was being transformed into a boomtown, the boys watched in fascination as new buildings, many of them shacks, were being thrown together to the constant sound of pounding hammers. The most impressive building was the state's capitol on the plaza. A Chinatown had sprung up nearby, and a strange new sound was the clanging of gongs summoning the workmen to meals. Never having seen a Chinese person before, the children were intrigued by unfamiliar looks and dress.

As they explored the town one day, the young Murphys witnessed a scene that to them seemed curious—a *Californio* custom that, before long, would disappear forever...it was wash day on the Guadalupe River. The children looked on wide-eyed as the local people arrived in their ox-pulled wooden *carretas* filled with laundry. After washing their brightly colored clothes and hanging them to dry on the bushes, it was time to start the *fandango*. A colorful scene emerged—the men in their short jackets of colored silk, white shirts with black scarves around their necks and pantaloons of velveteen, open from the knee down and ornamented with buttons and gold braid. Over their long black hair tied with ribbons in back, they wore black, shallow-crowned, wide-brimmed black hats.

Isaac Branham. The meetings of the first legislature met at his adobe in the pueblo. FROM *HISTORY OF SANTA CLARA COUNTY*, 1881.

State House Building, 1849. COURTESY ARBUCKLE COLLECTION.

The women were festively dressed in long, full skirts, blouses with short puffed embroidered sleeves; their hair braided, parted in the middle and tied with ribbons. A guitarist and a fiddler filled the air with lively music and, with a natural ebullience and gift for enjoying life, the *Californios* sang and danced. Then, when the stringed beef from the barbecue was ready, they feasted. The wide-eyed young Murphys gazed wistfully at the happy scene.

Nearby was the *plaza*, the heart and life of the town where everything from colorful religious processions to the bear and bull fights took place—the latter eventually outlawed by pressure from the padres at the *adobe* church of St. Joseph. With the influx of prospectors who stopped over to stock up on supplies on their way to the Mother Lode, San Jose was fast becoming a mercantile center. For lack of space some merchants stacked their goods in piles beside the entrance doors.

Social life abounded during the first legislative session with parties every night. The only Murphy to attend was John, a member of the first elected city council, with his bride-to-be, Virginia Reed, daughter of James Frazier Reed of the Donner Party. They made a handsome couple as they arrived at the beautifully furnished homes on the Plaza of Don Antonio Suñol and his wife, Doña Dolores Bernal, and of her sister, María Pilar Bernal and her husband, Don Antonio Pico.

Climaxing the social season was the Inaugural Ball on the top floor of the newly completed capitol. The elegant invitations, printed in gold on pink satin, featured the names of the committee which included Lt. Governor John McDougal, General Mariano Vallejo, Caius Tacitus (always called C.T.) Ryland, Governor Peter Burnett, James Frazier Reed, Joseph Aram, and over from Sacramento, John Bidwell and Captain John Sutter, who had been delegates to the Constitutional Convention in Monterey.

While the legislature was in session a quaint custom, also soon to disappear, took place. This was the manner in which an invitation to the ball was issued—another new experience for the children. Fascinated, they watched as a troubadour in festive costume mounted on a spirited horse circled the plaza strumming his guitar and singing a lively tune. When he had the attention of all he extended an invitation to the *fiesta*. After exchanging a few pleasantries he moved on to serenade each home with a special invitation.

At the formal affair Governor and Mrs. Burnett received guests, assisted by their daughters, Rhea, who later married the younger Martin's attorney, William T. Wallace, and Letitia, who became the wife of attorney, banker and legislator C.T. Ryland.

The younger Martin had first met the governor at Sutter's Fort when he was straightening out Captain Sutter's business affairs by selling off some of his property. Martin bought two of the Sacramento lots. He and Mary were too occupied with their children and readying their new home at the ranch to attend social functions.

CHAPTER FOURTEEN

Bay View Ranch—1849

With Mary and the children settled in the Hernandez *adobe* in San José, the younger Martin had a list of unfinished business to take care of. Although he had a buyer for the Ernesto, the deal included three thousand head of cattle that had to be accounted for. The next project for Martin was to make arrangements for a home to be built on the *Borregas rancho*. Instead of the customary California *adobe*, he and Mary Bolger wanted a frame house. There were no sawmills in California at that time capable of turning out the lumber, but with his new affluence (he sold the Ernesto for fifty thousand dollars) Martin ordered a twenty-room New England style house, pre-cut and milled to his specifications at Bangor, Maine. Each piece was to be labeled and crated. He then made arrangements for it to be shipped around the Horn to the Embarcadero at Alviso and from there transported by wagon to the ranch.

He still had to find someone to assemble the pre-fabricated pieces after they arrived, and that would turn out to be something of a puzzle. He located a man named Dawson who agreed to do the job in exchange for five hundred acres of *Borregas* land. There was yet another complication. There were no nails to be had in the valley at that time. Fortunately Dawson was resourceful. He said he would put it together with pegs and tie the beams with rawhide. It turned out to be the first frame house in Santa Clara Valley.

With the arrangements completed, Martin now had time to take Mary and the children out to see where their new home would be. Bursting with excitement, they hiked over the hilly land that was so different from the flat Sacramento Valley. They finally decided upon a site for the house by a grove of oak trees, looking out to the bay with a view of the steamers making their way between San José and San Francisco.

Martin Murphy, Jr. COURTESY RUTH MURPHY POLK COLLECTION.

Mary Bolger Murphy. COURTESY RUTH MURPHY POLK COLLECTION.

Bay View is the 20-room home Martin had pre-cut and milled to his specifications and shipped around the horn to Alviso. No nails were available in the valley at that time, but a man named Dawson agreed to put it together with pegs and tie the beams with leather. It was the first frame house in Santa Clara Valley. COURTESY SUNNYVALE HISTORICAL SOCIETY AND MUSEUM ASSOCIATION.

Martin now had a question, "What are we going to name our new home?" Mary Bolger said she thought that the land grant name *Pastoria de las Borregas* was too long and difficult to pronounce. After discussing several names, they all agreed on Bay View. Even though it was a pretty name, the locals would always refer to it as the Murphy Ranch and later, with the advent of the railroad, as Murphy Station.

As they tramped around the property, nature-loving Mary was elated to see wild roses; valleys dense with ferns, and lakes and streams bordered with willow and sycamore trees, whose leaves, rippling in the cool breeze, were reflected in pools of water. She dreamed aloud of the garden she would plant. Martin talked about the rich soil and the planting he would do of grain and hay, and orchards too, with cuttings of pear trees from the Santa Clara mission orchards which had been treated so ruthlessly by squatters.

Their conversation was interrupted by the clip clop of horses' hooves and they looked up in time to see a stagecoach, pulled by six mustangs, racing down nearby El Camino Real which ran between the site of the house and the bay. A few months later they would cheer to hear the stage driver shouting, "California's been admitted to the Union," and to recognize Governor Burnett sitting beside him waving his stovepipe hat. The governor had been in San Francisco the day before when the mail steamer *Oregon* arrived bearing the news that Congress had passed a bill on September 9, making California the thirty-first state.

The younger Martin, assisted by his three oldest sons, James, Martin and Patrick, soon started to prepare the land for planting. Never one to waste time, it wasn't long before Martin moved in cattle for grazing.

Although the boys were good workers, the pursuit of their education was a top priority with Martin and Mary who had both been deprived of education in Ireland. At first they hired a private tutor, but in late 1849 the opportunity for formal education arrived with a pair of Italian Jesuits, Michael Accolti and John

Nobili, who had sailed down the Columbia River and then down the Pacific Coast on a lumber boat to San Francisco. They had been working with the famous missionary priest, Pierre de Smet in the Oregon Territory.

California was ready and eager for a Jesuit college. Both William Hartnell of Monterey and Dr. John McLaughlin of the Hudson's Bay Company had strongly urged the Jesuits to establish schools in California. But it was the determined personality of education-minded Father Accolti, scion of an aristocratic Roman family, combined with the equally persistent Martin, that brought about the establishment of Santa Clara College, the first institution of higher learning in California. In 1912 it would achieve university status.

Father Accolti deluged the Jesuit General in Rome with letters and moved among the people of the state soliciting support for a college. When at last he did hear from the general, instead of a reply to his letters, the missive, dated two years before, announced Accolti's appointment as superior of the Oregon missions. Although disappointed that he would be unable to see the college become a reality, he sailed back to Oregon satisfied with the knowledge he would now be in a position to lend the school his support by way of sending qualified teachers.

His traveling companion and good friend, Father Nobili, not yet imbued with the same zeal for founding the college, stayed on for a while as assistant pastor of St. Joseph's Church in San Jose. A frail man, he was still suffering from the debilitating strain of working with the traders, trappers and Indians in the wilds of New Caledonia. He frequently lacked for food during this adventure. At one point he was reduced to eating moss. Even so, when the cholera epidemic hit San José in 1850, Father Nobili worked night and day nursing the victims.

The Murphys escaped the dreaded disease, but were shocked to learn that Dr. John and Elizabeth Townsend, leading members of the Murphy-Stephens wagon party, were fatal victims. Townsend had treated a cholera victim, contracted the disease himself and passed it on to his wife. They left a small child, John, in the custody of his uncle, Moses Schallenberger.

Because of the meager accommodations for the priests at the adobe church of St. Joseph, Father Nobili accepted the invitation of the pious Don Luís Peralta to stay at his home. It was here that he first met the younger Martin, which led to a meaningful friendship for the two men. Unfortunately, Nobili's relationship with Padre José María Pinyero, pastor of St. Joseph church, was not a harmonious one. Pinyero, possibly influenced by the book, *The Wandering Jew* by Eugene Sue, and related anti-clericalism of the time, was prejudiced against the Jesuits. In a petition to the town council he opposed the establishing of the college—a proposal that failed.

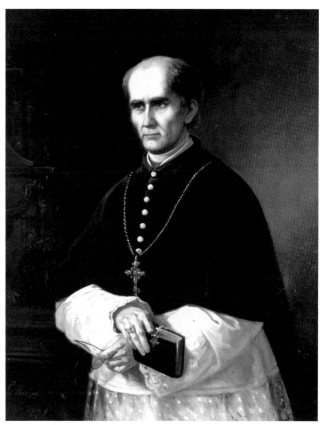

Bishop Alemany, of the Dominican order, was appalled to find education in California lacking. COURTESY SANTA CLARA UNIVERSITY ARCHIVES.

More good news for Martin in his quest for an education for his children was the arrival at the end of 1850 of Bishop Joseph Sadoc Alemany. The Spanish-born priest who was an American citizen and had been serving as Dominican provincial to the United States was appointed by the Pope to be bishop of Monterey and California. He faced a formidable task. Not only was the Church in California in disarray, handicapped by a shortage of priests, but it suffered an appalling lack of educational facilities.

Understandably, he strongly supported the opening of a college at Mission Santa Clara and encouraged the Jesuits to take it over. The Murphys were at Mass in the mission one Sunday, when the bishop announced the appointment of Father John Nobili as pastor of the mission church and president of the new college. His official letter was dated March 4, 1851. In the bishop's travels around his large diocese on horseback, by stage or boat, he frequently spent the night at Bay View. In fact, one room, called the Bishop's Room, was set aside for him.

By this time Father Nobili had become an enthusiastic supporter of the school. He entered his new assignment with dedication, as he faced the responsibility of restoring the mission buildings which had been badly damaged not only during the occupation of General Castro and his men but by squatters and by the Americans who stayed there during the Mexican War. Father Accolti had aptly described the mission buildings in a letter to the younger Martin as "plundered and reduced to the condition of a big stable."

Letter from Bishop Joseph S. Alemany to Father John Nobili appointing him pastor of Mission Santa Clara and the church of St. Joseph, March 4, 1851. COURTESY SANTA CLARA UNIVERSITY ARCHIVES.

In a letter Bishop Alemany wrote to the Jesuit provincial in Rome, he described the buildings as "wretched victims of negligence." Unfortunately, the mission's departing pastor, Padre Suarez Real, had more worldly interests and less concern for the mission properties during his seven-year pastorate. He not only scandalized the local people by living with a common-law wife, but sold off lots and buildings for his own gain. After he received his orders to return to Mexico, he detoured by way of the gold fields where he had great success. In Mexico he became involved in politics and left the order—no doubt by request.

Following the Mexican War many of the American occupants of the mission properties felt that the properties belonged to the United States and refused to leave. Father Nobili wrote to the younger Martin saying that the few buildings attached to the mission that had not had been either, "sold, bestowed or filched away, were in a condition of dismal nakedness and ruin."

Appreciating the priest's predicament, Martin purchased the California Hotel on adjoining property and deeded it to Father Nobili. Another legal entanglement involved the occupation of part of the priests' residence by the former major-domo of the mission, James Alexander Forbes, and his large family. It was settled in exchange for tuition for his sons in the school and other considerations.

This was just the beginning: the ecclesiastical hold on the property was shaky. Claims were made to everything but the mission church itself, and most of Father Nobili's time was spent establishing title to lots and adobe buildings scattered around the central mission and restoring them for classrooms. One purchase that didn't set well with the younger Martin was the fourteen hundred dollars paid to his brother, John Murphy, for title to a piece of the orchard land he had bought from an earlier claimant. Two other Americans, Judge Joshua Redman and Thomas O. Larkin, claimed title to mission land on the basis of a grant from Governor Pico in 1846.

In response to a request from the bishop, Father Accolti in Oregon sent four Jesuit instructors to assist Father Nobili. They included Peter de Vos who enjoyed a reunion with his old friend and with Governor Burnett, a strong supporter of the school. While both were still in Oregon, Father de Vos had received Burnett into the Catholic Church. The saintly Jesuit also became a close friend of the elder Martin Murphy.

In May of 1851 the college opened with twelve students. They included the younger Martin's sons, James and Martin; Governor Burnett's son, John; James Alexander Forbes's son, Miguel, and José Antonio, the son of Don Antonio Suñol. That same year the California Wesleyan (University of the Pacific) obtained a charter, but classes didn't start until a year later. A friendly relationship developed between the two colleges. On the occasion of Pacific's first graduation ceremony Father Accolti was invited to sit on the platform.

Santa Clara College in 1854. COURTESY SANTA CLARA UNIVERSITY ARCHIVES.

Notre Dame College Music Hall. Notre Dame was reported to be the most beautiful school for young women west of the Mississippi and the first chartered woman's college in the west. It received strong financial support from both Martin, Sr. and Jr. FROM *LIGHT IN THE VALLEY* BY MARY DOMINICA MCNAMEE, S.N.D.DEN., 1967.

Children of early pioneers from all over the state soon came to attend the Jesuit school, including the sons of Abel Stearns of Santa Barbara, Job F. Dye, William M. Keith and Alpheus Thompson. Many

Californio names such as Alviso, Pinyero, Berreyessa, Vallejo, Estudillo and Bandini also appeared on early rosters. For some of the Spanish-speaking students bilingual classes were necessary.

In 1852, a year after entering college, Martin and Mary's son, James, was the victim of an undiagnosed disease. That same day while he lay dying in an upstairs bedroom, a continuance of life took place downstairs as Mary Bolger gave birth to a baby boy. They named him James Thomas after his deceased brother. Martin and Mary had already known the sorrow of losing three very young children, and now, their eldest son. Yet in her great faith she was able to say, "The Lord giveth and the Lord taketh." They consoled themselves with the thought that they had twenty years of happy memories of James. In a gesture of kindness to Mrs. Murphy and in appreciation for her husband's generous gifts to the college, James had the distinction of being buried in the mission church—the only non-religious ever to be so honored.

During the formative period of Santa Clara College, the Murphys, with three daughters still to be educated, played an important part in the establishment of a school for girls. In the spring of 1851, at a meeting in San Francisco with Bishop Alemany, Martin met, by chance, two Belgian Notre Dame nuns, Sister Loyola and Sister Catherine Marie. They had just arrived from the Oregon Territory to await the arrival of four nuns from Cincinnati who were traveling by way of the Isthmus of Panama. One of the nuns, Sister Aloysius, would join the faculty of Notre Dame College.

The Notre Dame nuns had been invited by a priest to open a school in San Francisco, but on their arrival they learned that another priest had already brought in the Daughters of Charity of St. Vincent de Paul. There wasn't room in San Francisco at that time for two Catholic schools. The bishop told the nuns he thought San Jose would be an ideal location for a girls' school. The younger Martin, who was present at the meeting, could not contain his enthusiasm. He offered to conduct them to San Jose so they could look over the area and invited them to stay at Bay View Ranch. On their trip by wagon down El Camino Real from San Francisco, the sisters gazed admiringly at the hillsides carpeted with green and awed at the fields of brilliant yellow mustard growing ten feet tall.

The next day, after spending the night at Bay View, the sisters traveled by wagon to San Jose where they experienced a culture shock. The once quiet little Spanish village had become an American boomtown and not a very attractive one. As Martin and Mary drove them around the visitors saw cattle roaming the rutted, muddy streets fronting meager adobe houses, tents and flimsy wooden shacks that had been thrown together to accommodate the representatives attending the meeting of the first legislature that was already in session.

The sisters' lack of enthusiasm was apparent on their faces. Not even the warmth and congeniality of Mary Bolger could lift the spirits of Sister Catherine. Cautiously, Sister Loyola, who favored the move to California, asked her what she thought. Unable to contain her outraged feelings, the usually quiet sister exploded in French, thinking, of course, that the Murphys could not understand her. She stopped abruptly when she saw the amused smiles on their faces. They explained to the embarrassed sisters that they had lived in Quebec and were conversant in French.

The next day the world took on a brighter look. The sisters had a happy reunion over breakfast at Mission Santa Clara with Father Nobili. Seven years before, they had shared some pretty overwhelming experiences with him on an eight-month voyage aboard a two-masted brig from Antwerp around the Horn to the Oregon Territory. Their spirits buoyed, the sisters announced that they were prepared to meet the challenge.

Returning to Bay View, the nuns were surprised to learn that Bishop Alemany had arrived. He came to celebrate Palm Sunday Mass at the mission and to organize the parishes of Santa Clara and San Jose under the Jesuits. When the sisters arrived for Mass they were taken aback by the combined sounds of the ringing of the church bells, the singing of the Indian choir and the beating of drums, punctuated by exploding fireworks.

The next day they had an opportunity to meet members of the local families. They included the younger Martin's sister, Margaret, and her husband, Thomas Kell, who invited them to stay at their home on the Almaden Road until they became settled. As work on the school progressed, the sisters moved into town with the Charles Whites until Don Antonio Suñol and his wife, Doña María Dolores Bernal, invited them to stay at their *adobe* home on the *plaza* where the sisters could tutor their daughters and give them piano lessons.

The younger Martin lost no time in organizing a group of supporters called the Senators who were eager for their daughters to have a convent education. They

assisted the nuns in choosing a piece of property, bordered by willow trees, that adjoined that of Don Luís Peralta. They wisely observed the importance of water from the adjoining acequía.

Mayor Charles White and his wife, Ellen and the younger Martin deeded parcels of their land to them without charge. Levi Goodrich, their neighbor on the left who became the architect for the school buildings, soon followed suit giving them his two adjoining lots.

The arrangements for the building fund started off well. The two Martins led off with pledges of $600 followed by $800 from the Senators. Sister Catherine, the business member of the partnership, became concerned when some of the Senators' pledges could not be honored. The elder Martin offered to go security for any debts the sisters would be obliged to contract in the building of the school and came to the sisters' rescue more than once. With the drought of 1850-51, even though he could not market his cattle, he borrowed two thousand dollars on his ranch which he loaned to them at no interest.

The sisters also had a housing problem that was soon settled. Father Nobili arranged with a Belgian countryman of theirs, Dr. Caneghan, for them to stay in a little unoccupied hospital until their first buildings were completed. The good doctor was guardian of three-year-old Johnny Townsend, and at his request, the sisters took him in as their first boarder, accompanied by his Irish nurse and an Indian servant girl. Moses Schallenberger, the little boy's uncle and administrator of the Townsend estate, compensated them well. At the age of seven Johnny was moved to Santa Clara College and placed in the care of the Jesuit fathers.

The life of the much-admired Father Nobili was to come to an abrupt end in 1856, just a year after the college received its charter. While supervising the construction of a student chapel, he stepped on a nail. The wound at first seemed trivial, so he paid little heed. Unfortunately lockjaw set in and he lived only two weeks longer.

The prospect of a school for their daughters was a big boost to the local residents—in fact to prominent families all over the state. It was the first women's college west of the Mississippi and, in time, with its landscaped grounds which featured fountains, arbors, sculptures and plantings to set off superb architecture, Santa Clara College was regarded as the most beautiful college west of the Mississippi.

Although the education of his children was now settled, the younger Martin never lost interest in expanding his land holdings. Early on he purchased additional acreage on the *Pastoria de las Borregas* from Mariano Castro. In 1853, after leasing eight hundred acres of grazing land from his neighbor, Juana Briones, on her adjoining *Rancho La Purisima Concepcion*, he purchased three thousand acres from her to give to his daughter Lizzie Yuba and William Taaffe for a wedding gift.

The Widow Briones, as she was called, was one of California's most famous women. North Beach was known for many years as *Playa de Juana Briones*. Noted for her nursing skills, she took in sailors off the ships who were suffering from smallpox, scurvy or any other illness, and nursed them back to health using Indian herbal remedies she had learned from her mother. The only payment she asked was the satisfaction of seeing her patients recover.

The younger Martin, an admirer and good friend of Briones, never ceased to marvel at her energy and ability as she farmed her land, raised cattle and periodically made trips to San Francisco in her *carreta* loaded with hides to exchange for merchandise. She still found time to nurse the sick and went into Purisima's neighboring villages to treat the Indians.

In 1860 the younger Martin decided Bay View's land was too valuable for grazing cattle so he looked to the south and San Luis Obispo County. He purchased the seventeen thousand acre *Rancho Santa Margarita* of Don Joaquin Estrada who was having problems establishing boundaries and followed in 1864 with acquisitions of the four thousand acre *Atascadero*, the thirty-nine thousand acre *Asuncion* and the twelve thousand acre *Cojo rancho* at *Point Concepcion* in Santa Barbara County. The Murphy properties, covering over seventy thousand acres and described by one writer as a "principality," now extended for fifty miles along the base of the Coast Range.

Meanwhile, all was moving along well at Bay View. Martin's grain crops thrived as did his grapevines. Even the gardens Mary had anticipated with such pleasure were growing well—one plant that grew to gigantic proportions was the fig crossed with a Banyan tree which had been transplanted from Mission Santa Clara. Like the Banyan it continued to drop its roots into the ground so that eventually it could accommodate eighty people under its branches.

Forward-looking Martin became the first in the valley to import farm machinery, which he had shipped by way of the Isthmus of Panama. He was the first to fence in his ranch and the first to import Norman horses from France. With an Irish love of horses, one of Martin's favorite entertainments was to have horse races on the ranch.

Among other "firsts," he leased Sacramento land to the California Stage Company for the first terminal in the Sacramento Valley. In 1858 he built the first brick building in San José, which was used for commerce, a courthouse and a post office. The younger Martin's lending practices were extraordinary. He charged one and one-half percent interest when the standard rate was three percent.

In an 1872 story in San Francisco's *Morning Call*, the writer, who had been making a survey of the wealthiest men in the state, wrote, "We begin with one who, in our humble opinion, with the exception of Stanford, Hayward, Sharon, or Parrott, when all is told, is the richest man in California—Martin Murphy of the Santa Clara Valley.

"The exact value of this man's real estate and personal property we are unable to estimate. The leagues and leagues of valuable lands scattered through the counties of Santa Clara, San Luis Obispo, Santa Barbara, Kern, Fresno and Tulare, swarming with bands of horses and herds of wild cattle, do not constitute the bulk of Martin Murphy's wealth. His railroad shares, the hundreds of thousands in investments equally safe, the blocks of elegant buildings in San Francisco, San Jose, San Leandro and other towns, together with his ready money, must carry his wealth close into the double file of millions."

"Mr. Murphy is well known in California, and though he keeps aloof from the whirl of fashionable life, still he is ever welcomed and esteemed in the highest circles of society." This description was evidence of the low-key Murphy life style. Their home was a haven of Irish hospitality. In *The Journals of Alfred Doten*, the author tells of playing fiddle for a party at Bay View. The Murphy boys had come for him. "We had a very jolly time—cotillions, reels, waltzes, jogs, polkas and other figures without number were danced. Mrs. Murphy and Mrs. Lacy, with two gentlemen, showed us how the Irish jig was performed...Irish whiskey was plenty and all went in freely for having a good time.

"Little attention was paid to time or regularity in the dancing, and it was a regular Irish scrape throughout. About 5:30 in the morning the ball broke up... I went upstairs and turned in for an hour, after which I got breakfast... Mr. Murphy gave me ten dollars for my services."

No one who came to Bay View's door was ever turned away. In his dictation to the Bancroft Library, Bernard Murphy told about his mother—that she welcomed all with a meal. The strong point of her character was honesty. If a horse or a cow was being sold she would tell the buyer every thing about the animal—more than he ever wanted to know. She was economical, but never parsimonious—believed that every good thing that came along was a dispensation of providence.

One of those who stopped at Bay View on his way walking to San José from San Francisco, was an Easterner named Henry Willard Coe, who later became a prominent citizen of the valley. After first acquiring one hundred sixty acres in the "Willows" he bought *Rancho San Felipe*. According to his grandson, who was named for him, he used to enjoy telling how the Murphys on that particular occasion not only fed him but provided him with a horse and saddle which he explained by saying, "That was the way they did things."

Typical of so many of the Irish, who had been deprived of the right to vote or a seat in Parliament before coming to the United States, the Murphys became active in politics. In 1853, the location was Bay View when the Murphy family participated in the organizing of the Democratic Party. John Murphy was elected secretary. The younger Martin's home served as a meeting place until a permanent location in town could be found. Although the younger Martin never ran for public office, he was loyal to the party, as were his brothers who were elected to several posts.

Jim, a farmer and the least politically minded, served on Santa Clara County's first grand jury. John gained a seat on the first San Jose City Council, and subsequently, a half dozen or more public offices. Equally active was their younger brother, Barney, who served four and one-half terms as mayor and in the state assembly. Over the years, within the walls of Bay View, could be heard talk of Bull Run, Gettysburg, of San Juan Hill and Manila Bay.

CHAPTER FIFTEEN

Martin, Sr.—Family Ties 1850

At the age of sixty-five the elder Martin had suffered enough from the fleas in the Hernandez adobe which, typical of its time, had been built of mustard stalks, willow branches and mud. He was now comfortably settled in the spacious hacienda on his *Rancho Ojo de Agua de la Coche* built of redwood from his steam-powered sawmill in the nearby mountains.

California in 1850 was the right place and the right time for this Irish patriot who possessed great energy and undaunted courage. Land was cheap, and he had prospered during the Gold Rush from selling his cattle and wheat to the prospectors who came to California from all over the world, enabling him to purchase land grant *ranchos*. Because his integrity was as well known as his hospitality he gave public service in the early '50s as justice of the peace for the Burnett Township.

In choosing the site for his new home he took advantage of the view he had so admired when five years earlier he rode down the valley below the Pueblo de San José with Sutter's army in the Micheltorena Campaign. The site looked across a broad expanse of level, oak-dotted land to a conical peak named by the Spaniards, *El Toro*. Wayfarers on El Camino Real, the most traveled road in California at that time, called it Murphy's Peak for almost fifty years.

In this earthly paradise, as he liked to call California, he would sometimes reflect on the road he had traveled with his family—from the trying times in Ireland, to Canada, and then over rivers, lakes and canals to Missouri. He rejoiced at the memory of Father Hoecken, the Jesuit missionary who suggested that he take his family to California, and over their long journey by wagon train to become the first overland party to scale the summit of the Sierra Nevada and make wagon tracks in California.

Martin Murphy, Sr. COURTESY BETH WYMAN.

With Martin and Mary Bolger and family now settled on the *Pastoria de las Borregas rancho*, Bay View, the patriarch had much of his family nearby. An exception was his daughter, Mary Miller, who was living with her husband, James, and their large and growing family in San Rafael. The distance separating them was no problem for this sturdy pioneer. He frequently rode his pony the eighty or so miles each way for visits.

Living in the area was his eldest daughter, Margaret, and her husband, Thomas Kell, who had come overland from Quebec in 1846 with their four children and were established on a piece of the *Ojo de Coche rancho* land which the elder Martin had given them. His second daughter, Johanna Fitzgerald, having lost her husband, was arriving soon with her six children from Quebec, escorted at her father's request by her brother, Bernard.

Music of romance filled the air at *Ojo de Coche* in 1850. Within a little more than a year all four of the elder Martin's single children would make trips to the altar—John with Virginia Reed of Donner Party fame, Helen with Charles Weber, founder of Stockton and Dan with Mary Fisher, daughter of Captain William Fisher of the nearby *Laguna Seca* ranch. Bernard, who while living with his father on the ranch had little opportunity to meet young women, would surprise everyone by bringing home from Quebec an Irish bride, appropriately a native of Wexford.

Yet to be heard from was his second son, Jim, who had abandoned his thriving lumber business in San Francisco when his workers took off for the Mother Lode. After having a try at mining it wasn't long before he arrived at the home of his father. Around the dinner table one night he told the family that gold digging was not for him—that he had a family to take care of. His brother, Dan, told him he had just heard that the *San Francisco de las Llagas Rancho* was for sale. When Jim showed interest, the elder Martin suggested that they buy it together.

The twenty-two thousand-acre spread bordered his father's ranch on the south. Bernard, had already acquired the twelve thousand-acre *Las Uvas rancho* to the west and the four thousand-acre *La Polka*, bordering the *Llagas* to the east. Added to this, the elder Martin acquired a portion of the *Las Animas Rancho* for his daughter, Johanna Fitzgerald. The Murphy family now owned close to fifty thousand acres of ranch land in southern Santa Clara County.

Noted for his hospitality, the elder Martin kept his door always open to travelers who would spend a night or so on their journey between San Jose and Monterey. In fact early maps of the period noted Murphy's Rancho (sic). Many of his guests later wrote of his kindness, of his soft Irish brogue and his warm welcome. Still, it was his youngest daughter Helen, the pretty, animated, convent-educated young woman who had crossed the plains with the wagon party on horseback and on foot, who caught their fancy. A young Lt. William Tecumseh Sherman, later a general in the Union Army, wrote of her as "a sweet and attractive girl, bright and witty as a sunbeam." He told Dan Murphy

that army men on their trips inland arranged their plans in order to stop at the Murphy *rancho.*

Bayard Taylor, a reporter for the *New York Tribune* on his way to Monterey on foot to cover the Constitutional Convention, reported in his book, *El Dorado or Adventures in the Path of Empire*, of stopping at the Murphy ranch, "a well known and welcome resting place to all Americans in the country," and of his meeting Helen and other guests, a Mr. Ruckel of San Francisco and a Mr. Everett of New York. In the course of conversation, when Taylor expressed an interest in *El Toro*, Martin invited him to ride up the next day to see it.

Taylor wrote that Mr. Murphy called for two horses from his *caballada* to be saddled and that they rode across the west side of the valley to the foot of the mountain. They then followed a winding cattle trail, up rocky ridges through patches of stunted oak and chaparral to the summit where they had a commanding view of the sixty-mile long valley. He described the mountain as "rising like an island in the sea of air so far from all it overlooked," and the range upon range of coastal mountains parted by deep, wild valleys, in which they could trace the course of streams shaded by pine and redwood. To the north, at their feet, lay the valley of San Jose.

When they returned to the house they found that another guest had arrived, Father Dowiat, a Jesuit missionary from Oregon. Martin was interested to learn about the Indian massacre that took place the winter before. The priest said he had been there the day it occurred. He held them fascinated with his description of the tragic event—adding that he had assisted in interring the bodies of Dr. Whitman and his fellow victims.

Meanwhile, if the elder Martin had yearnings for his native Ireland, he could visit with John Tennant, a native of County Wexford who built his Twenty-One Mile House, a stagecoach stop and a hotel with a barroom, on nearby El Camino Real. It was the first in this part of the state. The first building effort of this former piano tuner was called Tennant's Seven-Mile House, located at what would later be called Edenvale at the start of the Monterey Highway leading from San Jose to Monterey. Down the road, across from the Eighteen-Mile House, the Murphys had their warehouses.

Of Martin Murphy, Bancroft wrote: "Surrounded by his children, Murphy prospered, living in patriarchal abundance, his herds, his lands and his numerous households."

CHAPTER SIXTEEN
Bernard, the Quiet One

Word came in 1851 of the death in Quebec of Patrick Fitzgerald, husband of the patriarch's daughter, Johanna. Concerned for her welfare and that of her six children, the elder Martin asked his son Bernard to make the long sea voyage to Canada to bring them to California. Bernard was the quiet one, lacking the more flamboyant personalities of his younger brothers, John and Dan, though equally shrewd in purchasing land.

On the trip across the plains, while his younger brothers were looking for excitement, Bernard elected to stay close to his father. After a brief stint in the gold fields, he bought the twelve thousand-acre *Rancho Las Uvas* that bordered the *Ojo de Coche* on the west from Maria del Carmen Berreyessa, widow of Lorenzo Pineda who received the grant from Governor Alvarado.

Bernard was happy to make the trip back to Quebec. Aside from the fact that he could never refuse his father, it gave him an opportunity to marry the girl he had left behind. Distance had made communication with her difficult. Living on the ranch he had little opportunity to meet any of the few available women in California at that time. He arrived at his destination only to learn that his girl had given up on his ever returning and married another. All was not lost, however, for he met and married a comely, bright Irish lass from County Wexford named Catherine O'Toole.

In addition to his bride, his sister, Johanna, and the six young Fitzgeralds, he added to his entourage Mary Bolger Murphy's sister, Elizabeth, her husband, John Sinnott, and their four children. They traveled by ship to Chagres where they commenced the hot, humid journey across the Isthmus by mule and canoe.

At the same time, traveling the same route from New York on the *Empire City*, (afterward a warship in the Civil War) was a young Irish priest, Eugene O'Connell who was destined to become one of Northern California's outstanding bishops. A friend of the Murphys, he occasionally celebrated Mass at their Bay View ranch. He was escorting two Dominican nuns en route to Monterey, where their order was to start a school, and four Notre Dame sisters on their way to San Francisco to join Sister Loyola and Sister Marie Catherine.

The trip turned into a harrowing experience for the nuns. At the entrance to Chagres, after they boarded a boat manned by Indians, a group on shore shouted a warning to them not to land, that murderers were lying in wait for them. Suddenly they saw ten naked Indians rowing towards them armed with knives and daggers. To their rescue came a Jewish doctor and several companions armed with pistols. They had just boarded their next boat when they passed a number of murdered bodies floating by, victims of the same men who had almost attacked them. As if that trauma weren't enough, when their steamer arrived at Gorgona, it went aground in deep sand, causing a delay of three days.

For their trip through the jungle the sisters doffed their religious garb for night robes—loose Mother Hubbards of violet calico, and huge white sunbonnets such as they wore when hanging up clothes at the convent. They were a colorful sight as they plodded along on their mules through the dense jungle growth, dodging branches that hung low over the trails, making it necessary at times to bend over parallel with their mounts. Lagging behind, absorbing the scenery, was gentle Sister Aloysius. Suddenly her mule shied at a puddle of water and just as he shot off into the dense chaparral she took hold of an overhanging branch. As she hung there, in horror of falling to the ground, with perfect timing along came Bernard Murphy. He gently lifted her down before going in search of her recalcitrant mule, which he found leisurely browsing in the brush. Her fellow sisters teased Sister Aloysius, comparing her to Absalom in the second book of Samuel. They said, "You fared better than he because he hung by his hair when he got caught in the branches of a tree."

During the excitement of their Chagres adventure the sisters were dreaming all the while of the Indian children they expected to have for their charges in the Willamette Valley of Oregon. What they didn't know was that at that very same time in San Jose, Sister Loyola and Sister Catherine were meeting with Bernard's brother, the younger Martin. They were laying out plans for their new school in San Jose where the Fitzgerald and the Sinnott girls would one day become students, two of whom would join the Notre Dame order. Annie Fitzgerald became Sister Anna Rafael, one of Notre Dame's prestigious nuns as both scientist and poet. Mary Sinnott became Sister Mary Anselm. Meanwhile, the quiet, reserved Bernard could bask in the sun of his rescue experience, a scenario worthy of his more adventurous brothers, John and Dan.

The party sailed from Panama on the *Sarah Sands*. When the group arrived in San Francisco, Martin was there with his wagon to meet them. Mary Bolger had come along—eager to see her twin sister, Elizabeth. They enjoyed a joyous laughter-filled-with-tears reunion. Martin was happy to see his cousin John Sinnott whom his father had brought from Ireland to Canada. John farmed a parcel of Martin's land at Mountain View for several years before buying farmland on the *Rancho Milpitas*.

After safely delivering his charges, Bernard moved his bride, Catherine, into the tin house he had ordered shipped around the Horn and transported by wagon from Alviso to his four thousand acre *Rancho La Polka*, part of the *San Ysidro rancho* he had purchased from Isabel Ortega. Her sister, María Clara Ortega lived with husband, John Gilroy, on Maria's share of the *San Ysidro rancho*. Gilroy was a Scotsman who, while still in his teens, jumped ship at Monterey to become the first non-Hispanic citizen of California. The city of Gilroy bears his name.

Bernard and Catherine's marriage was a happy but short one. It ended abruptly in 1853. Bernard was a passenger on the *Jenny Lind*, a side-wheel steamboat en route from Alviso to San Francisco when the boilers exploded. Bernard was killed along with thirty others, including his nephew, Thomas Kell, Jr., son of his eldest sister Margaret.

The year following the tragic accident, the patriarch, after years of riding his pony twenty miles to San Jose to attend Sunday Mass, gave Archbishop Joseph Alemany four acres of land on the Llagas *rancho* for a church and cemetery. On it the elder Martin built a twenty by forty-foot chapel which he named San Martín for his patron saint, Saint Martin of Tours. Archbishop Alemany came down from San Francisco to bless the little church, and at the same time bless the re-burial sites of Bernard, and his nephew, Thomas Kell whose bodies had been moved from a San José cemetery. Attending the ceremony were the elder Martin, with Bernard's widow, Catherine O'Toole; Margaret Kell with her husband, Tom; Dan Murphy with his wife, Mary Fisher; and Johanna Fitzgerald. In time numerous members of the John Gilroy family would also be interred in the little cemetery.

After the railroad arrived in 1869, a station was built called Mil's Switch, around which the town of San Martín was laid out in 1892.

Bernard and Catherine O'Toole Murphy had one son whom they named Martin John Charles Murphy for his grandfather. He was the first child baptized in the San Martín chapel. From Stockton, to become the infant's godparents, came Helen and Charles Weber. The patriarch's good friend, Father Peter de Vos, performed the ceremony. After that, Father de Vos, who had served in the Napoleonic Wars before going to New Orleans and crossing the plains with the noted frontier priest Peter de Smet, came every third Sunday from San Jose to say Mass.

The much-loved young Martin died in 1872 of an undiagnosed illness while a student at Georgetown University in Virginia.

The explosion of the steamboat Jenny Lind. FROM *CALIFORNIA CAVALIER, THE JOURNAL OF CAPTAIN THOMAS FALLON,* EDITED BY THOMAS MCENERY, 1978.

CHAPTER SEVENTEEN

John, the Adventurous One

John Murphy has been described as the "most reckless, conscience-free, devil-may-care member of the Murphy family." He loved adventure and he loved excitement. His life was filled with plenty of both. He was still in his teens when he crossed the plains with the wagon train headed by his father and Elisha Stephens. When there were streams to ford John was there. When there were buffalo to be had, he hunted.

Although he had had risky near-encounters with Sioux warriors, his closest call with eternity came later in the trip. He was among the party of six including his brother Dan, his sister Helen, Elizabeth Townsend and two hired men, which separated from the larger group to ride ahead on horseback to John Sutter's fort. Should it become necessary, they would return to the wagon group with provisions. While the main group took a tributary of the Truckee, the riders set off following its main stream passing to a large lake that would later become known as Lake Tahoe. They crossed the river and followed the west shore of the lake for a piece, rode over the mountain to the Rubicon and reached the American River which turned out to be even more winding than the Truckee, making it necessary to cross and crisscross against a strong current.

Along the way John escaped almost certain death when his pony stumbled on a rock and John was swept into the water, dashing him against boulders and making him too weak to swim. Helen and Dan, joined by the others, were terrified as they ran downstream, expecting to find their brother's mangled body on the shore. To their great relief they found him grasping the branch of a willow tree with a life or death hold. As they pulled him from the water he lost consciousness and it took several days for him to recover.

John Murphy, the fun-loving, more adventurous son of Martin, Sr. COURTESY *CARTE DE VISITE* BY WRIGHT, 1992 COPY BY DANIEL KASSER & BYRON WOLFE.

The party of six arrived at New Helvetia and barely had time to rejoice over their reunion with the rest of the wagon party when they learned of the Micheltorena Rebellion. The patriarch told his two younger sons what Sutter had told him about foreigners not being welcome and how they had all decided to join Captain Sutter's army. So, off to war rode John with his father and four brothers. It wasn't much of a war. Their biggest excitement came when John and his brother Dan were taken prisoner by General Castro while on reconnaissance with a dozen other soldiers near Buenaventura. They were released four or five days later after they promised to stop fighting against the insurgents and to encourage others to stop. An explanation was never recorded, but some way or other John lost his trousers and was reported riding back to New Helvetia with a blanket across his front. Dan was wearing shoes that were threadbare. It had been a year since they left Irish Grove, Missouri, and the shoes simply wore out.

John and Dan went first to Sutter's Fort and then up to the cabin in the mountains to assist in the rescue of the snowbound women and children. Later, Dan took a job working cattle at the French ranch and Martin, Jr. purchased the *Rancho Ernesto* on the Cosumnes, eighteen miles from New Helvetia while the rest of the family continued on to the pueblo. There, John took a job clerking in the mercantile business of Charles Maria Weber, who became a trusted friend of the family. It was through Weber that Martin, Sr. was able to acquire the *Ojo de Agua de la Coche rancho*.

Hardly a year had passed since the Micheltorena affair before the U. S. declared war on Mexico. During the Bear Flag Revolt at Sonoma General Mariano Guadalupe Vallejo was apprehended at his home on the *plaza* at Sonoma by John Fremont's men and taken to Sutter's Fort where he was held prisoner.

It didn't take long for John Murphy to become involved. When U.S. Consul Oliver Larkin, a friend of the general, learned of Vallejo's arrest, he asked John to deliver a letter to him. Ironically, Don Mariano was an admirer of the Americans and favored California becoming a territory of the United States. On receiving the letter, the general asked the engaging twenty-two-year-old Murphy if he would wait while he wrote a letter to take back to Larkin. John readily agreed. In Monterey the consul acted quickly in securing a release for Vallejo from Commodore Sloat. Sloat was reached just in time as he was on his way out to his ship which would take him to the East Coast. Larkin again asked John to serve as courier.

Departing at dawn he covered the one hundred-twenty-mile stretch to Yerba Buena under a penetrating July sun without stopping, except to change horses. After delivering the dispatch to Captain Montgomery he continued on his way to the fort, detouring by way of Sonoma to impart the good news to Señora Vallejo. He then escorted the general, who was now ill with malaria, back to Sonoma.

Thomas Oliver Larkin was the only U.S. Consul to California. When he learned of General Vallejo's detainment at Sutter's Fort he hired John Murphy to deliver a letter to the Mexican general. FROM *THREE YEARS IN CALIFORNIA* BY REV. WALTER COLTON, U.S.N., 1851.

By this time the well-liked Vallejo and the personable young Irishman had become good friends, but their paths did not cross again until 1886 at a fortieth anniversary celebration of the raising of the American flag in Monterey. The occasion also marked the seventy-eighth birthday of the courtly general whose once vast Sonoma estate was now reduced to a few acres and whose vast herds of cattle to one milk cow. Yet he performed with aplomb the honor of first raising and lowering the tri-color flag of Mexico before raising the Stars and Stripes on the same flagstaff used by Commodore Sloat on July 7, 1846. While so doing, he said,

"This flag will float here forever." The ceremony completed, his face lighted up as he spotted his friend, John Murphy. After warm *abrazos*, he said he was happier over seeing John again than for any of the accolades he received that day.

Back in the *pueblo*, because of concern for the safety of the foreigners, Charles Weber was commissioned a captain, and John his first lieutenant, with instructions to enlist volunteers in the San José Militia. After Stockton and Fremont took Los Angeles, the war was believed to be over until the *Californios* rebelled at the treatment they received under U.S. Captain Gillespie and his drunken, badly behaved soldiers. Thoroughly incensed, José María Flores led the *Californios* to recapture Los Angeles.

John Fremont set off by ship with his men to join Stockton at the *pueblo* of Los Angeles. They were only a short time out when they met the *Vandalia* and learned of William Mervine's defeat and that all the Americans' horses had been taken. Without horses, Fremont and his men would be without hope. He ordered the ship to return to Monterey and immediately sent an urgent appeal to Weber. Sensing Fremont's desperation, Weber, usually joined by John, raided the *ranchos* for horses. When Don Francisco Sanchez, a *ranchero* with large holdings on the San Francisco peninsula, became incensed at the treatment the *Californios* received from the Americans, he gathered two hundred or so volunteers who were ready to rebel. This led to the bloodless "Battle of Santa Clara", also known facetiously as "The Battle of the Mustard Stalks" in which Weber and John were leading participants. This was the only confrontation of the war fought in Northern California.

On January 13, 1847, three days after Commodore Stockton raised the American flag over Los Angeles, Colonel John C. Fremont signed the Treaty of Capitulations of Cahuenga with Andres Pico marking the end of the Mexican War. At this point John was ready to take off in another direction—a political career. In the *pueblo*'s first municipal election in 1847 he was one of six *regidores* elected to assist the *alcalde*, James Weeks, in governing the *pueblo*. Three of the new men were *Californios*: José Noriega, Dolores Pacheco and Salvador Castro. Joining John were Americans James Frazier Reed (his future father-in-law), and Thomas Campbell.

The trials of war were now behind them, but the calm following the storm was soon to end and with an event that would shake the whole world. In January of 1848 John Marshall found the famous gold nuggets at Sutter's Mill. At this point John Murphy put politics on hold to rush to the gold fields and eventually was joined by just about everyone else in the *pueblo*. He and his brother Dan partnered in Captain Charles Weber's Stockton Mining Company. They started mining at Weber's Creek and proceeded to pan and dig in streams at Sutter Creek and the Mokelumne River. At Weberville, John ran a trading post for Weber.

After Weber disbanded the group the two brothers continued on, arriving at Coyote Creek in Calaveras County. An Indian directed them to a flat by the creek where there was a rich gold strike. By this time they had decided there was more profit in dealing with the Indians so they set up a trading tent on the flat. After a succession of names, including Murphy's New Diggins, Murphy's Flat, Murphy's Rich Diggins and Murphy's Camp, it finally, became simply Murphys. The village of that name lives on in the Sierra Nevada foothills. The Murphys had a good rapport with the Indians, trading *ponchos*, blankets, red shirts, bright cloth and trinkets for gold. Although during this period gold was valued by the United States Mint in Philadelphia at from sixteen to twenty dollars per troy ounce, the rate in the Sierra was $6 or whatever the item traded was worth to the buyer.

In 1858 the *San Andreas Independent* wrote that an Indian was said to have brought in a five-pound nugget to Murphy's establishment for which he received a blanket. The story lacks credence because John left Murphys at the end of 1849. Every night John butchered two bullocks for the Indian workers but he had one inviolable rule: no alcohol on the premises. By the end of 1848 Dan told John he had decided to leave. With his share of the earnings he said he wanted to start acquiring land and cattle.

In mid-1849 a distinguished visitor, Rev. Walter Colton, on a prospecting trip to the gold fields, visited the Murphy camp. He had served as chaplain on the U.S. frigate, *Congress*, when it arrived at Monterey. Commodore Stockton appointed him the town's first *alcalde*. He then had the distinction of publishing (with Robert Semple) California's first newspaper, *the Californian*. In his journal, *Three Years in California*, he described his visit to Murphy's camp and John Murphy inviting him into his tent.

Colton also described the tribe of wild Indians who gathered gold for John and how they respected his person and property. This, he said, could be attributed in part to the fact that he had married Pokela, the sister of José Jesus, the chief. He described her as "a young woman of many personal attractions and full of that warm wild love which makes her the Haide of the woods. She is the queen of the tribe and walks among them with the air of one on whom authority sets as a native grace—a charm which all feel, and of which she seems the least conscious."

John stayed on for another year. Finally, the time came to bid farewell to his lovely wife—a decision he had anguished over for some time. "My father is getting old" he said, "He needs me...life is so different there. You wouldn't be happy." He knew he could not assimilate into the Miwok way of life though he had submitted to the tribe's style of marriage ceremony in which his bride was carried to his tent by a strong brave, followed by the singing and dancing of joyous members of the tribe. As he departed in December of 1849, his heart was as heavy as the seventeen hides of gold carried on the backs of seven mules. It was more gold than that possessed by any man in California, but his thoughts were on Pokela.

John didn't know it at the time, but as he was leaving Murphy's Camp danger lurked around the corner. Fortunately, his timing was right. Two days after his departure the infamous Joaquín Murieta arrived in the neighborhood with his gang of *bandidos*. They robbed and killed two miners who were on their way home with their newly found riches and threw them into a prospector's hole.

Murieta had come from his native Sonora, Mexico with a caravan shortly after the discovery of gold. He found a strong sentiment among the Americans against foreigners, especially those who spoke Spanish. They were run out of mining camps, sometimes killed by vigilante committees.

The placer mines at Murphys were the richest in the Calaveras region with the possible exception of Dry Creek. All this gold, which came to John so easily, filtered through his fingers just as freely in the form of loans to his gambling friends who seldom, if ever, repaid them. His own gambling debts and his habit of never turning anyone down who asked for a hand-out were also part of the problem. San Jose attorney Frederick Hall wrote of Murphy in his 1871 book, *The History of San Jose and Surroundings*, that he was "generous, benevolent and had he kept his fortune he would have been one of the wealthiest men in the nation. But his bump of generosity is too great; he was too benevolent and the fortunes came too easy to be appreciated."

Back in San José, John met and fell in love with sixteen-year-old Virginia Reed of the Donner Party. He was the first of the patriarch's children to be married in California. The marriage didn't come off without a hitch; in fact they were married twice. First, James Frazier Reed, stepfather of the bride, was adamantly opposed. Then John's father, with his deep Catholic faith, was upset when they married outside the church. He insisted on Catholic nuptials. John's sixteen-year-old bride was a young lady of strong convictions. During the tragic snowbound episode at Donner Lake it was she, who, after her father had been banished from the party for killing a man during an altercation, had gone out into the night to bring him food, a shotgun, ammunition and a horse.

As Chester Lyman, a surveyor and an ordained Congregational minister, told the story, he was at Governor Burnett's when he was called out by Senator David Douglas who told him that John Murphy had requested that he marry him to Virginia Reed. Lyman said that he refused, having heard that Reed threatened to kill Murphy if he married his daughter. The couple spent the night at the mission. All was made right with John's father when they were married by Father Nobili at St. Joseph's Church. Virginia, who spent much time with the Breens in their cabin at the lake, had been so impressed with their faith she vowed if she ever got out alive she would become a Catholic. As for Reed, after Virginia and John's first child was born, grievances were forgotten.

After John settled down to married life with his beloved Virginia, he opened a mercantile business in San José. That same year John became involved in politics. Knowing all the old *Californio* families, the Suñols, the Picos, the Bernals and the Castros, and with a fluency in Spanish, he was elected the first treasurer and tax collector of Santa Clara County. It was a post he really didn't want, but one nobody else would take.

The first thing he needed was a repository for the official funds. When he learned that Judge Joshua Redman had an old iron safe he had salvaged from a whaling ship that had gone aground on the shore of Monterey Bay, he approached the judge's son, known as Red. He dangled before young Red's eyes the position of deputy treasurer at a salary of $250 per month if he could obtain the use of his father's safe. The deal was made. In Major Horace Bell's book, *On the Old West Coast*, he told of Red's eye-opening experiences—of how the treasurer would bring in one of his pals and say, "Look here, Red, my friend Jim has been in bad luck. He's been bucking monte and has gone flat broke. Give him a stake out of the safe, that's a good boy, and he'll bring it back to you in a day or two. His luck is bound to change." Red would give Jim or whomever a thousand dollars of the tax money and off he would go to break the bank. If he did he would bring back the money, otherwise, he never heard from him. One day he found it necessary to inform the treasurer Murphy that the iron safe was about twenty-five thousand dollars short, and the tax money wasn't coming in. This gave John food for serious thought. According to Red, "he dug down into his jeans... made up the losses of his gambling friends to the people of the county then and there and started a life of financial reform," adding, "for be it known the Murphys of San Jose have always had a reputation for strict integrity. Generous and careless at times perhaps, but always making good in the pinch."

Collecting taxes was not to the liking of John Murphy. In the fall of 1850 an election day was scheduled for Gilroy Township. He devised a plan whereby he would kill two birds with one stone. John rode down to the polling place, the *Rancho San Ysidro* of John Gilroy and, finding all the taxpayers present at one place at one time, thought he would get the tax collection over quickly. By the time he and Red reached the polling place, all the ballots had been cast in a china teapot that served as ballot box on a long table in the sala of the Gilroy home. Two clerks and two judges sat at the table with Don Juan Gilroy, as the locals called him, acting as inspector. The well-known and well-liked Scotsman was the first non-Hispanic resident of California, having skipped ship in Monterey in 1814. He married María Clara Ortega, daughter of Don Ygnacio Ortega and they inherited a portion of the large Spanish land grant, *Rancho San Ysidro*. The city of Gilroy bears his name.

Not only were all forty-five of the voters present but Indians, Mexicans and *vaqueros* of mixed blood had come from surrounding *ranchos* and were milling around, jingling their spurs. Some were squatting on hides or blankets playing Monte. But most were hoping for a little excitement when a couple of well-dressed strangers with a city look, definitely not from this part of the country, arrived, saying they were on their way to Monterey on business. John and Red recognized one of them as an athlete they had seen in a foot race at Sacramento and the other as his manager. They remembered he was a world champion and known as the fastest man on foot. Obviously there was a motive to their visit and they soon made it known.

The strangers had heard of Nicodemus, one of John Gilroy's many sons. He was famed not only for his horsemanship but also for his speed in running down wild cattle. They also knew the Gilroys were proud of their son's agility and they assumed the couple was wealthy and wouldn't hesitate to bet on him. That would also go for all those gathered around. John Murphy was concerned—took the runner's manager aside and asked him to identify his champion to the unsuspecting crowd. He received a cool rebuff for his effort. Just before the race was to start, Señora Gilroy came running from the house with a red bandana jingling with $600 in coin to bet on her Nicodemus. John was again rebuffed when he told her who the runner was and tried to persuade her not to bet on Nicodemus. Unable to change her mind, John did what he thought was the next best thing. He took $600 of the taxpayer's money and bet it on the champion with the intention of giving it to Doña María Clara if her son lost and he was sure he would.

Red and John were the last to leave the Gilroy house. As they walked toward the race site, he noted, out of the corner of his eye, Grove Cook, a mountaineer and member of the Democratic Party going in the direction of the house. He thought no more of it as they joined the crowd. It seemed like the whole countryside had gathered around in a fever pitch of excitement. As the gun went off to signal the start of the race, the professional runner practically flew across the starting line. Suddenly, Nicodemus woke up to what was happening. He swept past his opponent and, as he crossed the finish line, did a handspring for the cheering crowd. An exuberant Señora Gilroy collected her winnings little realizing the gallantry of John Murphy who had tried to protect her from loss.

Still in a state of euphoria over their local boy making good, the spectators slowly moved back to the Gilroy house for the counting of the ballots. It was a foregone conclusion that the vote would be predominantly Whig. To the amazement of John and Red who were strong Democrats as each vote was read it turned out to be for the Democrat. Only one person in the room seemed not to be surprised. It was Grove Cook—while everyone was watching the race, he had stuffed the ballot box with Democratic votes. Stashed away in his pocket were the Whig votes. The election officials refused to swear to the returns, canceling the total vote of the district. But Grove Cook had had his fun.

During the fifties Virginia kept busy having babies while John continued to be elected to political offices. After serving his term as treasurer, in 1851 he was elected county recorder as well as alderman for the now incorporated city of San José. People continued, however, to call it the *Pueblo* for several years. Winning elections was never any problem for John. In 1853 he was again elected county recorder and to the city council. He also played an important role organizing the San Jose Democratic Party which was roundly supported by all the Murphys for at least the rest of the century. Dr. A.J. Spencer was elected president and John and Samuel Morrison, secretaries. In 1854, John, the only member returning to the city council, was elected its president. Midway through his term he resigned but returned to office the following year. In 1856 he was the only holdover on the council. When Mayor Lawrence Archer resigned John succeeded him, but his term of office as mayor was short for he soon after resigned. He apparently had had enough of the city council because the following year he ran for county sheriff and, as usual, was elected.

One of the first projects he became involved in was the building of a road across the Santa Cruz Mountains. There had been only Indian paths linking the Santa Cruz mission and Mission Santa Clara. Father Fermín Francisco de Lasuen, in 1771, returning from Santa Cruz wrote, "I returned to Santa Clara by another way—rougher but shorter and more direct. I had the Indians improve the road. "Around 1850 Zachariah "Buffalo" Jones and Mountain Charlie McKiernan established toll roads on the summit.

Stockholders from Santa Cruz and Santa Clara counties formed the Santa Cruz Gap Joint Stock Company in 1856 with Adolph Pfister as president. Also part of the company were three "viewers" or surveyors for laying out the road on the Santa Cruz side and three on the Santa Clara side—one of whom was Sheriff John Murphy.

Politics wasn't John's only forte. A social animal, John helped organize and became an officer in the Young Men's Social Club. He served on the committee for the Washington Birth-Night Ball at the capitol and, when the Santa Clara Agricultural Society gave a ball at city hall, John was listed on the program as a floor manager. His handsome, young man-about-town nephew Patrick Murphy was on the reception committee. Among the popular young society matrons listed were Virginia Reed Murphy and Camilla Horn,

mother of Gertrude Atherton who became a well-known novelist.

Back to the election board—in 1859 John was re-elected sheriff. The life of a lawman was never dull. He captured a dangerous *desperado* named Pancho Daniel and incarcerated him. The wily Pancho managed to escape—causing no end of excitement. While the townspeople looked on, shaking in horror, Sheriff Murphy took off after him through the *plaza* followed by the constable, the marshal and the deputy marshal. Deputy Marshal Jake Miller took aim and fired, breaking Pancho's arm and killing his horse. The *bandido* turned on Miller and had him on the run—at which point Pancho transferred his saddle and bridle to Miller's horse and took off. He hid in a haystack and talked an Indian into bringing him food. When John got word of this he bribed the Indian to reveal Pancho's whereabouts. He demanded Pancho come out, threatening to burn the haystack. The desperado, knowing he was cornered, gave himself up. The townspeople were greatly relieved when Sheriff Murphy set off with his captive to deliver him to the sheriff in Los Angeles.

More sensational, though, was John's experience involving a squatter's dispute that led to three deaths. It occurred on the ranch of one Henry Seale whose brother had decided to eject a squatter named Paul Shore from Henry's property near Mayfield. During a dispute between the two men Henry's neighbor S.J. Crosby arrived, allegedly to return a pistol he had borrowed from Seale. A story circulated that Crosby had set Seale's dog on the squatter, urging the animal on until Shore ceased to breathe. Thomas Seale, who had taken it upon himself to eject the squatter, turned himself in to Sheriff Murphy, stating the action had been in self-defense and asking for a public investigation.

The next day, when John received word that a plan was afoot to lynch Crosby, he sent his under-sheriff to arrest Crosby, charging him with being an accessory to the killing. After a preliminary examination Crosby was released but when Thomas Seale's trial came up he was summoned as a witness. Crosby was fired at and killed on the street near the courthouse by Thomas Shore, brother of the victim. A spectator at the trial who had opened the door to see what was going on also received a fatal shot, bringing the total to three dead over the squatter's incident.

The Settlers' War of 1861 had all the color and excitement of an old Western movie, excepting that not a shot was heard. It was another squatter-inspired theme. Antonio Chaboya received the U.S. patent to his twenty-five thousand-acre *Rancho Yerba Buena*, a beautiful piece of land that extended from the eastern hills to Coyote Creek. Unable to persuade the squatters who had moved in on his land to leave, he called for help from Sheriff John Murphy who, in turn, called for a posse. One thousand men turned out in front of City Hall, where from the balcony he called the roll. When he asked them if they were armed and ready to assist him, the crowd yelled—NO!!!! He then asked them if they would arm themselves and go out with him to the *rancho*. They again yelled—NO!!! He then said he had tried to follow the law, and they refused—that there was nothing more he could do, so he dismissed them. The crowd headed for the Evergreen Schoolhouse on the *Yerba Buena rancho*. By this time the number had grown to two thousand—all armed with sidearms, scythes, two cannon, and several flags. As described in *The Journals of Alfred Doten*, "an Ambro type of man was mounted on top of an omnibus with his camera trying to take a view of the scene."

At one o'clock they started for town. The lead wagon carried the band followed by all the wagons with their flags flying, and eighty-three carriages carrying the ladies. In the first buggy was Mr. Ballard with his lady holding a rifle. Next came about one thousand horsemen each with a rifle or shotgun on his shoulder riding four abreast. After entering through First Street they marched through all the principal streets and out onto Washington Square. Sheriff Murphy, riding alone, drove into the center of the square. The crowd moved in to hear his speech. As reported by Doten, John Murphy said he was their sheriff and had never before failed to do his duty, but they had prevented him this time—that he derived his power from the people, but the people had taken the power away, so what could he do? They cheered him heavily from time to time, gave three groans for the land sharks and three more for lawyer Matthews—"jolly good groans."

The commander, Ari Hopper, dismissed them, saying they must always stand ready to turn out in any such emergency. As the crowd dispersed, all grew calm and quiet. Chaboya and his neighbors were, in the course of time, able to work out a peaceful settlement.

That same year the Civil War broke out. Although California was loyal to the Union cause, some of the Southerners expressed strong sentiments. With California's extensive shoreline, it behooved the residents to form military companies in the event of foreign complications. Few saw action in the major conflict. John was appointed captain of the Johnson Guard, an unattached unit with a first and two second lieutenants. By this time, with his growing family, John decided to build a larger home. Virginia, a loving wife and mother, was active in St. Joseph's Church and as their daughters entered Notre Dame Academy she became a close friend of John's niece, Sister Anna Raphael. When John decided to build a new home for Virginia his father gave him a lot on South Second Street which the *alcalde* of San Jose had granted to him. Nothing was too good for John's Virginia. He had the lumber shipped around the Horn and paid carpenters eighteen dollars a day to construct the spacious house the outstanding feature of which was a wide veranda across the front, providing a view of the rolling eastern foothills. Located next door to the Trinity Church rectory, the property, running south for almost a half square mile, had a fine vineyard and an orchard of pear and apple trees. The theater on South First Street backed up to the Murphy property and when the weather was rainy John, always sensitive to Virginia's well being and concerned that her slippers might get muddy, carried her in his arms out the back door to the theater.

Retiring from political life after his stints as sheriff, John opened a dry goods business followed by a real estate and insurance office. Due to his liberality as well as his weakness for games of chance his fortune was depleted but, as was aptly said, "...of which he took little notice." At his death in 1892 his wife Virginia took over his insurance business and was the first woman fire insurance writer in California.

CHAPTER EIGHTEEN

Dan Murphy: The Cattle King

With his unlimited vitality and vision, Dan Murphy, the youngest of the elder Martin's five sons, was a natural leader and seemingly born to succeed. He inherited these characteristics from his father, who had not only shown leadership with the Murphy-Stephens party, but was looked up to as the Irish chieftain in Canada and again in Holt County, Missouri, where many of his countrymen followed him.

Daniel Murphy. Although only seventeen years old when the Murphys crossed the plains, he seemed always to emerge a leader. Early on he became interested in cattle raising, becoming known as the Cattle King of Nevada. At his death in 1882 his obituary read that Daniel Murphy had more cattle and more land (including his four million-acre ranch in Durango, Mexico) than anyone in the country. COURTESY JOYCE HUNTER.

As a teenager on the wagon train crossing to California, with his energy and drive, Dan seemed always to be in the middle of things. Moses Schallenberger wrote in his well-known account of the Murphy-Stephens party of Dan's leading a party of six in search of cattle that had wandered away during the night from their camp on the Green River. Because of Dan's quick thinking, Moses wrote, they had averted an encounter with the Sioux who were staging a war party.

Hardly had they recovered from that close call but they were surrounded by another two hundred Indians. Dan later said he was inwardly shaking but decided to put on a bold front. He signaled to the strangers to dismount. They did. In fact they turned out to be friendly members of the Snake tribe, looking for the same Sioux warriors Dan's group had seen the day before.

On yet another occasion Dan took the lead with a party of six which had separated from the main wagon train to ride ahead by horseback to Sutter's Fort. In the event the wagon party should run into trouble they could send help. En route Dan, moving ahead of the rest to check out the terrain, became the first white man to step on the shores of what would become known as Lake Tahoe.

After they returned to Sutter's Fort following the Micheltorena Rebellion, Dan was offered a job working on the *rancho* of Charles Weber's partner, Guillermo Gulnac, at the old French Camp on the San Joaquin River. It was there he got his first taste of cattle ranching, a taste he would savor and one that would ultimately lead to his becoming one of the largest cattlemen in the country, annually shipping six thousand head out of Halleck, Nevada.

He and a man named John Williams cared for Gulnac's seventy head of cattle belonging to James Lindsay who was later killed by Indians. Although the Murphys had experienced good relations with the Indians this was remote country. In a land grant case Dan testified in a deposition to the large number of hostile Indians in the area. The threat of their presence no doubt contributed to his decision to leave to see his father at his newly acquired *Ojo de Agua de la Coche Rancho*.

The following year the United States and Mexico were at war. It was thought to be over after Commodore Stockton and Lieutenant John Charles Fremont raised the United States flag at the *Pueblo de Los Angeles*. It wasn't long, however, before word came that fighting had resumed—that the Californios had recaptured Los Angeles. When Fremont learned that the Americans had no horses, which, of course, was a serious situation, he sent out an urgent plea to Charles Weber for horses and volunteers.

Dan soon became involved in leading one of Weber's raiding parties at the *ranchos*. At Mission Dolores, he met Jacob Harlan and John Van Gorden who were on their way to the San José *pueblo* to secure housing for the Harlan-Young Party, one of many which arrived in 1846. Dan invited them to join him. Along the way they experienced typical California hospitality at the late Rafael Soto's *Rancho Rinconada del Arroyo de San Francisquito*. His widow of half a dozen or so years and her daughters welcomed them and prepared a meal of *chili colorado, tortillas* and *frijoles*. Afterward they danced with the men until two in the morning.

A short time later Dan enlisted with Fremont's California Volunteers. On their way to Monterey to join Lieutenant Colonel Fremont the Volunteers had an unexpected meeting with the enemy, led by Manuel Castro. During the ensuing encounter, called the *Battle of Natividad*, Dan lost his friend, Joseph Foster, a member of the Murphy-Stephens party.

Fremont acted quickly in moving his men to Mission San Juan Bautista before leaving south on what would be a long, miserable journey. As Dan later told his father, "Most of the time we sloshed through mud in heavy rain and cold."

Scarcely a year passed before gold fever seized the country. On the move again, Dan and his brother, John, got an early start in the gold race—first as members of Captain Weber's Stockton Mining Company and then, taking off on their own, by opening a merchandise tent at Coyote Creek in Calaveras County. Dan stayed at the diggings long enough to accumulate a golden nest egg before departing for home to assist his father in running the *Ojo de Coche rancho*. He now had the wherewithal to pursue his two major interests, land and cattle, and at each he would be successful beyond his wildest dreams.

By 1850, financially secure, Dan was pursuing the hand of fifteen-year-old Mary Fisher of the neighboring *Rancho Laguna Seca*. They made a stunning couple. Dan, tall, strongly built with dark skin, and a gray streak through his black hair, was said to be the most handsome man in California at the time with more than his share of charisma. Mary, a beauty in her own right, inherited her fair skin and blue eyes from her English father, Captain William Fisher, and her dark, almost black hair from her Spanish mother, Liberata Ceseña. They pronounced their vows before Father John Nobili at Mission Santa Clara in January of 1851.

William Fisher, skipper of a sailing ship called the *Maria Teresa* which he owned with a friend, Julian Hanks, sailed between Mazatlan, Mexico and Monterey, transporting merchandise for sale to the early California settlers. There were rumors of a bit of buccaneering, too—not uncommon for the time and place to avoid high custom duties. On a trip inland, Fisher fell in love with the Santa Clara Valley. When he heard of a land auction in Monterey he made haste to be there. He successfully outbid John Gilroy and Charles Weber, who had *ranchos* nearby, for the twenty thousand-acre *Rancho Laguna Seca*. The two men told Fisher they thought he was crazy to pay six thousand dollars for it when there was so much cheap land available.

Paying no heed to their advice, the English captain's first action was to hire Chester Lyman to accompany him on horseback from San Jose to make a survey of his newly acquired property. Afterward, they spent the night at the elder Martin's *rancho*. Lyman, an author (*Around the Horn to the Sandwich Islands and California*) as well a surveyor and later professor at Yale University, wrote of their visit—of the elder Martin's fine farm and of his daughter, Helen, serving them a supper of eggs, bread and milk.

Captain Fisher lived only five years after moving to California, not quite long enough to see his daughter married to the up and coming Dan Murphy. After his death his wife, Liberata, had a difficult time managing the farm with its acres of grain, a vineyard, horses, mules, oxen and fourteen hundred head of cattle—all this in addition to a sawmill and a mill on Canoas Creek, both of which he had been part owner.

Dan's cattle on Las Llagas Rancho. COURTESY HISTORY SAN JOSE.

Ivy Farm, Dan Murphy's home in San Martin. COURTESY JOYCE HUNTER.

Dan, especially knowledgeable about cattle, stepped in to help his mother-in-law manage her affairs and subsequently bought half the *Laguna Seca* from the Fisher family for twenty thousand dollars. This increased his holdings substantially having recently bought part of the *Ojo de Coche* from his father.

In 1853, Dan and Mary entertained a visitor from Germany, Charles Weber's brother, Adolph. He had come at the behest of his parents in Germany to report on Charles, whom they had not heard from since he left home fifteen years earlier. After his visit with Charles and Helen Murphy Weber in Stockton, he stopped in San José to call on Helen Weber's brother, John, and his wife, Virginia. Adolph had spent two days fishing and shooting gray squirrels with John when the elder Martin returned from a trip on his pony to San Rafael where he had visited his daughter, Mary, and her James Miller family.

In a letter to his parents, Adolph wrote "Mr. Murphy was very glad to see me and conversed with me in his Irish dialect. He is about sixty seven years old years old, rather thin and gray, but very healthy and strong."

The patriarch invited Adolph to accompany him three miles south of San Jose to the home of his eldest daughter Margaret and her husband, Tom Kell, where they spent the night. The next day the two men reached Dan's *San Martín rancho* where the elder Martin now lived and where Dan and his wife, Mary Fisher, gave Adolph a warm welcome. With a twinkle in his eye, Dan told Adolph he would introduce him to life on a California cattle ranch. This he did and for a member of a prominent, intellectual Bavarian family, it was quite an experience.

The following morning he was told there was to be a cattle sale on the Llagas—that two dealers were coming from San Francisco to buy for their butcher shops. The action began with the rounding up of six hundred head of wild cattle. They had been driven in a frantic chase from the hills and vales by fifteen Mexican, Indian and Chilean *vaqueros*, whose long black hair, hanging almost to their shoulders, and long, curling mustaches extending below their chins, added interest to the hectic scene.

Adolph was fascinated not only with the sight but also with the sound of the thundering herd—the mooing of the calves, the oxen and cows. Occasionally, when one broke loose he had to be chased, lassoed by a *vaquero*, and brought to the herd bucking violently. If a *vaquero* had a problem with an especially recalcitrant animal, he would lasso his hind legs, fasten the rope to the saddle horn and drag him to the appointed location.

The skill of the *vaqueros* amazed the German visitor. The elder Martin explained to him that in breaking colts, the emphasis on the training was for the safety of the *vaquero*. "It takes time and patience," he said, "and close observation of the good points of a horse to train him to a high degree of efficiency. He had to be swift of foot, obedient to the least touch of the reins, the pressure of the knees and the sway of the rider."

On hand for the sale, with their herds bearing their ranchos brands, were neighboring *rancheros* Agustin Bernal from the Santa Teresa, Pedro Chaboya from the Yerba Buena, John Gilroy from *Rancho San Ysidro* and Agustin Narvaez from the *Rancho San Juan Bautista*. The *vaqueros* moved slowly, separating the fattest animals. The price that day was fifty-five dollars per head. Dan told Adolph that the previous year he had bought one thousand head in Baja California at ten dollars each and, after fattening them on his rancho, sold them for fifty dollars per head.

The elder Martin was always the first in the household to get up in the morning and, after looking after his own business he, like the others, spent the whole day on his little pony. After Adolph's first day in the saddle he wrote home, "You can imagine how my center of gravity feels after spending the day on the hard, peculiar wooden Mexican saddle, as well all the other muscles after being in this unaccustomed position."

On another occasion Adolph joined Charles's brother-in-law, Tom Kell, in gathering up *vaqueros* to drive a herd of cattle to Dan's corral. Along the way they stopped at a creek to rest from the oppressive Indian summer sun. Instead of offering brandy and whiskey for lunch (breakfast and dinner would consist of beef steak, white bread, baked without yeast and tea) one of the *vaqueros* suddenly turned up with dried mustard stalks, an oak limb, and a rib of beef that he

had gotten from some Mexicans down the way who had butchered a cow. From some unknown source they produced salt, cayenne pepper and onions which they served with a shot of brandy as an hors d'oeuvre before the delicious beefsteak lunch.

Adolph never lacked for California hospitality. The elder Martin, ever the accommodating host, invited him for a ride through the western hills to see his steam-powered sawmill in a beautiful redwood grove. Afterward they visited Catherine O'Toole Murphy, the widow of Martin, Sr.'s son, Bernard, on the Polka Rancho. By this time Adolph was feeling more comfortable in the Mexican saddle and more familiar with the countryside, although he found it strange there were no fences and no roads other than the Monterey Road.

Riding one of his brother Charles's ponies which Dan cared for along with his own, Adolph climbed Murphy's Peak in the valley towards the west. "My little brown pony practically flew like a deer through the wild oats and over small creeks," he wrote. Another time Adolph took a sixteen-mile ride while Dan attended an auction with one of his colleagues and bought nine hundred and sixty head of cattle.

During the following three days, in a corral the cattle were branded and cutting marks made on their ears for identification. The scene was one of total chaos as the *vaqueros*, mostly Mexicans on this day, shouted their commands and curses in Spanish over the uproar of the bawling animals as they chased them from corner to corner in swirling dust. After being lassoed, the animal was taken to the fire to be branded…just another scene in the active life of a California *vaquero* for Adolph to witness first-hand.

By the end of October, it was time for Adolph to return to Stockton, but first he rode over to see Charles's *rancho*, *Canada de San Felipe y Las Animas*. The *rancho* was situated in a picturesque valley with about eighteen square miles of pasture where bear and other wild animals still roamed, and where, he learned, the United States Army corralled its horses during the Mexican War. On his last day Dan took him to the *Rancho Laguna Seca* so he could meet Mary's mother, Liberata Ceseña Fisher. Although the lovely looking Señora Fisher spoke no English, she greeted Adolph with the gracious manner so much identified Spanish California women.

Their roundabout route took them first north before turning west to the mountains behind the New Almaden quicksilver mines, which, Dan explained to Adolph, had played an important role during the Gold

Rush. The Fisher house was set back from the east side of the Monterey Road at the base of the hills. On the way they passed *La Laguna de Fisher*, a large pond where geese, swans and ducks and other wildlife congregated.

On his return to Stockton, Adolph wrote to his parents with news that must have disappointed them. He said that he had decided to stay in this country for a while. First, he explained, Charles was under severe stress in his efforts to obtain title to his land from the U.S. Land Claims Commission. Secondly, he felt he could save more money in one year in California than in three years as a lieutenant in the Bavarian Corps of Engineers. Charles, realizing his brother would now need a job, asked him to manage his *Rancho San Felipe y Las Animas*.

Ever moving ahead, when Dan learned of enormous grants of land being sold cheaply to Americans in Mexico he hired an agent who arranged for him to acquire a four million-acre *rancho* in Durango. On it rested a mountain of magnetic iron.

Dan would never be as politically active as his brother, John. Yet he was elected to the County Board of Supervisors in 1853 and to the Court of Sessions in 1855. In 1851 he was part of a group, led by Judge Davis Divine, promoting a railroad between San Francisco and San Jose. It was destined to become a reality (financed by another group of investors) only in 1864. There was still a need for interurban transportation. Finally, in 1868, Samuel Bishop, in a legislative action, declared Charles Silent, Daniel Murphy and himself associates to run and maintain a horse railroad within Santa Clara County.

The railroad ran between San Jose and Santa Clara along the famous Alameda, a road built some ninety years earlier and lined with willow trees from the Guadalupe River by a beloved Franciscan, Padre Magín Catalá, and his two hundred Indians. The same year the horse railroad opened, Dan was listed as a stockholder in the Bank of San Jose. He also managed to find time to serve as a director of the forerunner of PG&E, the only member listed as a capitalist.

Meanwhile, with his financial success, Dan built a spacious home in San José and another at San Martín where Mary gave birth to six children. Dan, who wanted children so badly, suffered deeply when they lost their first three in infancy. Then came their daughter, Molly, followed by Diana, whom he called Dannie because he had wanted a boy. Finally when their last child was a boy, they named him Daniel Mar-

tin for his father and grandfather. They were delighted when their daughter, Molly, married Daniel Chapman and produced two grandchildren. But their joy turned to sorrow when Molly died while the children were still small.

The horse railroad along the Alameda between Santa Clara and San Jose was run by Dan Murphy with associates Charles Silent and Samuel Bishop. COURTESY HISTORY SAN JOSE.

Dan indulged and lavished affection on his beloved Dannie. She grew into a great beauty with dark skin, black hair and violet eyes. Yet life in San Martín bored her, as did the two years she spent at Notre Dame Academy in San Jose—all the time distracted by dreams of a more glamorous life. Needless to say, the nuns were not unhappy to see her leave. Despite her lack of interest in the academic, she was not without talent—fluent not only in Spanish and French, but with ability at the piano and in singing.

Affluence was not the key to happiness for Dan and Mary. Their marriage encountered stormy seas. He was frequently away buying cattle and land, moving like a nomad from his *ranchos* in Santa Clara Valley to the San Joaquin Valley, Nevada, Arizona and Idaho. Their operation absorbed his time from dawn to dusk. Moreover it was rumored that he had a roving eye for the ladies. Mary filed for divorce, but they decided simply to live apart.

In 1872 Santa Clara County was beginning to close in on Dan. On his land and cattle-buying junkets he had become impressed with the wide, open spaces of Nevada. He decided on Elko County in northeastern Nevada as his base of operation, perhaps because it brought back memories of the Murphy-Stephens wagon train. Thirty years earlier, after traveling along the Humboldt River to the Sink, they had crossed through this country.

With his *vaqueros* he trailed twenty-five hundred head of cattle, crossing the Sierra Nevada at Bridge-port, California, then driving the herd east along the Humboldt River, turning at North Fork where he established his *Rancho Grande*. He enlarged his land holdings to include the Haystack, the Devil's Gate and the Halleck until he had seventy thousand acres in Nevada. He eventually spread over to the Bruneau country of Idaho, forming a partnership for the T Ranch with Jack and Lewis Byrnes and Barney Horne under the name Murphy-Byrnes and Horne.

During this period he participated in roundups on the Owyhee River, originally known to trappers as the Sandwich Islands River named for some Kanakas who were killed by Indians at the river's mouth. When the Sandwich Islands' name was changed to Hawaiian Islands, the river and the desert took the phonetic spelling Owyhee. Dan's brand was a triangle A with barbs and a Diamond A. The roundups on the sixty mile-long, thirty mile-wide Owyhee Desert, between the north and south branches of the Owyhee River, were on a large scale. All the wagons of each outfit camped within a mile of each other—an impressive sight—the long strings of cattle coming from every direction.

After the breakup of the partnership of the T Ranch, Dan's operations centered mainly along the north fork of the Humboldt. He had three or four different work groups—one crew took the herd to the railroad, another to the ranches and another worked on the ranches. Each crew consisted of the "cookie" with his chuck wagon, twenty-five to thirty Mexican buckaroos and a string of two hundred to three hundred horses.

Felipe Cariusa of Santa Barbara headed Dan's opera-tions. Felipe's wife would join him on occasion and bring her guitar. To Dan's immense pleasure she would dress in Spanish costume and put on a show for the outfit, playing guitar, singing and dancing. Dan was not an all-work and no-play person. He loved a good time. In a rather whimsical quote from the *Elko Independent* April 15, 1881 (it may have been a slow day for news) the writer said:

Dan Murphy, the well-known cattleman, arrived here from the West, by train last evening. He has been off rambling about in the Southern country during the winter...Dan has divided the time since he left here

between the senoritas of Sonora, the Gambusani of Arizona, and the refinements of Southern Texas, and now comes to enjoy the salubrious airs, Italian skies and bacon and beans of Eastern Nevada. Pending the replenishing of his cellar and larder at Halleck Station, he will take potluck with the boys in town.

From Halleck, Nevada, which became his headquar-ters during the shipping season, he yearly shipped six thousand head of cattle that had been driven in from White County, Idaho where he had a cattle partner-ship with A.C. Cleveland. Small wonder he became known not only as the Cattle King of Nevada, but the Cattle King of Idaho.

Dan's beloved Dannie came frequently to see her father. Their close relationship faltered on a visit in 1881 when she broke the news to him that she wanted to marry Hiram Morgan Hill. Dan was adamant. This goal-oriented man didn't consider Hill, a bank clerk (and there were rumors of a fast lifestyle), a proper choice for his daughter. The strong-willed Diana was just as adamant that she wanted to marry Morgan and so the decision was left in limbo.

When she returned to San José, she talked to her mother, who agreed with her father. In fact, Mary hired Peter Columbet, the personable scion of a prominent French family, to be her secretary in the hopes that he might be a distraction to Diana if she got to know him. But that didn't work. Without the knowledge or con-sent of her parents, Diana and Morgan were married secretly in San Francisco.

Within a few months Dan took ill with pneumonia during a wet snowstorm while trying to save his cattle that were to be loaded on the train at Halleck. When his condition grew worse he was moved to Elko, where he sent for his family. Diana and her mother arrived in time to see him at the hospital while he was still lucid. His last request of Diana was that she promise not to marry Morgan Hill. She agreed, but with feelings of guilt—she couldn't tell him they were already married.

Dan's funeral service was held in Mission Santa Clara, where his marriage to Mary Fisher had taken place thirty years earlier. Surrounded by family and friends, he was laid to rest in the Santa Clara Mission Cemetery in the Murphy family plot. According to press accounts his estate was estimated to be worth between fifteen and twenty million dollars. The *Elko Independent* of October 29, 1882, wrote in part:

Mr. Murphy was a man of remarkable social qualities and had endeared himself to a large circle of friends by his generosity, integrity in all business transactions, and his universal good nature, always having a good word for rich or poor alike. By the exercise of his sterling business qualifications he succeeded in amassing a princely fortune having immense herds of cattle in Nevada, California, New Mexico and Arizona being perhaps the largest landowner on the Pacific Coast.

From a San Jose newspaper:

He was a man of strict integrity; his probity of character, his veracity and honor were never doubted. His word was always as good as gold. In business circles he was known only to be respected and his unaffected manner endeared him to all his acquaintances...the 'Murphys' has become a family by-word and they might well be termed the Rothchilds of the West.

Dan's holographic will gives an insight to Dan Murphy, the man. The opening paragraph read:

In the name of God, Amen. I, Daniel Murphy, in the County of Santa Clara and State of California, in perfect health and memory, (God, be praised) do make and declare that this is my last will and testament in the manner following to say:
First: I commend my soul into the hands of God, my creator, hoping and assuredly believing through the merits of Jesus Christ my Savior, to be the partaker of life everlasting and my body to the earth whereof it is made.

Two or three years later Dan's widow, Mary Fisher Murphy, married Peter Columbet, the young man she had hoped would attract her daughter, Diana, and distract her from marrying Morgan Hill. Despite the difference in their ages the couple enjoyed a happy married life.

CHAPTER NINETEEN

Helen and Charles Weber of Stockton

Helen Murphy was the favorite of the family and of just about everyone else. General William Tecumseh Sherman, after a visit to the elder Martin's *Rancho Ojo de Agua de la Coche*, described her as "…bright and witty like a sunbeam." Her charms did not pass unnoticed by Charles María Weber either. When he was under house arrest at Sutter's Fort during the Micheltorena Rebellion, and she was there waiting the return of her father and brothers, they became well acquainted.

Weber had made a trip from the *Pueblo de* San José to New Helvetia to try to persuade his friend, Captain John Sutter, not to side with Micheltorena—that it was a losing cause. Sutter turned a deaf ear to his friend's advice and Weber, for his trouble, was placed under house arrest for the duration—later he would receive preferential treatment.

Helen Murphy Weber was the youngest daughter of Martin Murphy, Sr. and served as his hostess to travelers along the El Camino Real. Many wrote about her personal charm and attractive looks. COURTESY HELEN WEBER KENNEDY COLLECTION.

Charles Weber, a native of Bavaria and from a long line of Lutheran clergymen, married Helen Murphy. With her convent education in Quebec, she and Charles shared much in common. COURTESY HELEN WEBER KENNEDY COLLECTION.

With Helen at the fort was Elizabeth Townsend, whose husband, John, was serving as Captain Sutter's medical *aide-de-camp*. The two women had become good friends during their horseback ride from the Truckee River to Sutter's Fort with the party of six and Elizabeth often joined Charles and Helen in a game of whist.

During the long evenings Helen and Charles told each other of their experiences and how each finally arrived in California. The bright, dark-haired Helen, with her expressive gray eyes and soft Irish brogue, was a match for the sharp wit of Charles. Both spoke French fluently, had classical education—Helen at the Ursuline Convent in Quebec, where she was taught all the refinements of a young lady of that period. Charles came from a prominent intellectual family in Bavaria and a long line of Lutheran ministers.

Helen told Charles about Frampton—of their long trip over lakes, rivers and canals to Missouri. At this point her voice broke as she told of the death from malaria of her mother and her brother Martin's six-month old baby. Both occurred on the family farm on the Missouri River near St. Joseph. Elizabeth, sensing Helen's pain, quickly broke in to change the subject to the Murphy-Stephens wagon trip from Council Bluffs.

From time to time Pierson Reading, Sutter's chief clerk who was left in charge of the fort, would join in the conversation and, as couriers arrived from Sutter, he gave them reports on the progress of the war.

Charles described his childhood at Steinwenden in southern Germany toward the end of the Napoleonic period. He spoke of his father, who was one of seven generations of Lutheran ministers. With his father serving as superintendent of schools, Charles said he was assured the best possible education. In fact, so intense were his studies that he became ill and was unable to finish. This put an end to his dreams of a diplomatic career. Eager to see the New World, he decided to go to America to visit his uncle in Belleville, Illinois.

At the age of twenty-two, he arrived in New Orleans. "Imagine my surprise" he said, "to find the Mississippi River frozen over. So I worked at several jobs until I came down with yellow fever."

After recovering, he said, he became caught up in the border warfare with Mexico and moved over to Texas where he enlisted in the company of Sam Houston, adding that he fought in the Battle of Goose Creek which is commemorated by a small monument called Buffalo Bayou.

Finally on his way to visit his uncle, again he was sidetracked from his destination after reading glowing accounts in the local newspaper about California, written by John Marsh, a pseudo-medico and booster of California. The temptation was too much. He decided to see California first and signed up with the Bidwell-Bartleson party that was departing in a few weeks.

When the time came to leave in May 1841, the party's numbers had diminished. Wisely, he said, they joined with a band of Jesuit priests headed by Father Pierre Jean de Smet en route to Oregon Territory to work with the Flathead Indians. Their guide was the highly touted Thomas "Broken Hand" Fitzpatrick, a man with twenty year's experience in the West and well acquainted with the Indians.

Knowing that Helen was a Catholic, Charles made a point of telling her how he had become friends with the famous frontier priest, and that when the two groups parted at Soda Springs Father De Smet gave him a rosary which he still had. His group headed south and west over the great salt desert and the Sierra Nevada. After a short stop at John Marsh's *Rancho Pulpones* on Mount Diablo, Charles moved on to Sutter's Fort at New Helvetia where he presented letters of introduction to John Sutter.

He stayed on to work there for a while and was so impressed with what Sutter had accomplished that he dreamed of establishing his own city at a slough on the San Joaquin River. His voice rose with excitement as he told of the forty-nine thousand acre land grant, *El Campo de los Franceses* (French Camp) in which he was a silent partner with William Gulnac. It reminded him, he said, of his native Germany, with its tall tules, sloughs, fertile farmland, rivers and streams —and all within view of the mountains.

Helen and Charles had become well acquainted by the time the Murphy men returned from the Micheltorena Rebellion and headed for the camp in the mountains to rescue the women and children. On their return, Helen left for the *pueblo* with her father and her brothers, Bernard and John. Dan had decided to stay on in the area to work for Weber's partner, William Gulnac, looking after his cattle.

Charles returned to his store in the *pueblo* to catch up on his affairs which he had left in the hands of Frank Lightstone, a German to whom he later gave the business as a wedding gift. His many enterprises included a blacksmith shop and a flourmill, the first of its kind, which he built with Gulnac. He started a bakery, a manufacturing shop for candles, one for boots and shoes and a hotel called Weber House that had a saloon. In addition to all this, he traded flour, crackers and salt from his salt works on the east side of San Francisco Bay with the merchant ships called Boston ships that came to the *Embarcadero de Santa Clara* to exchange finished goods for products from his store.

By this time Helen had lit a spark in Charles whose glow, it seemed, would never grow dim. During the next five years he played the role of ardent suitor, well aware that she was one of the most eligible women in California. At the same time he became a good friend of the Murphys, making it possible for the elder Martin to purchase the beautiful nine thousand acre *Ojo de Agua de la Coche rancho* off El Camino Real which he had admired so much on his tour with Sutter's army. The beautiful stretch of level land, scattered with oak trees, extended into the hills on both sides of the valley.

Charles, who had become a Mexican citizen in January of 1844, was able to purchase the property which he, in turn, resold to the elder Martin (in the name of his son Bernard) for the same price. He hired Helen's brother, the friendly, outgoing John Murphy, to work in his store in the *pueblo* while the ever-loyal Bernard stayed on at the *rancho* to help his father with the cattle and the grain crop. Helen served as housekeeper and hostess to her father's many visitors, travelers on El Camino Real between the *pueblo* and Monterey. It was the busiest stretch of roadway in the state at that time. Helen's intelligence and Irish gift for words obviously impressed their guests, for many years later they wrote about her beauty and engaging manner.

During the Mexican War, Edwin Bryant stopped at the *Ojo de Coche*, as the *rancho* was usually called, with Weber, who was moving a *caballada* of three hundred horses from his *Rancho Laguna Seca* to *San Juan Bautista*.

He later wrote, "Mr. Murphy is the father of a large and respectable family who emigrated to this country some three or four years since to the United States, being originally from Canada. His daughter, Miss Helen, who did the honors of the rude cabin, in manners, conversation and personal charms, would grace any drawing room."

After Helen and Charles were married and living in Stockton, journalist Bayard Taylor, reporting for Horace Greeley of the *New York Tribune*, wrote of his earlier visit, "...I remember the day when hot, hungry and foot-sore I limped up to the door of her father's rancho in the valley of San Jose and found her reading a poem of mine (no author ever had a more welcome introduction). Her father saddled his horse and rode with me to the top of a mountain, and with her own hands prepared the grateful supper and breakfast which gave me the strength for the tramp to Monterey."

Yet another traveler who chronicled his visit with the Murphys was Chester Lyman. He had served two years in Hawaii as a missionary before coming to California where he worked as a surveyor. In his book, *Around the Horn to the Sandwich Islands and California*, he told about his Santa Clara Valley experiences and of stopping at the elder Murphy's house. In his diary he wrote, "The family received us kindly, the old man being a clever, generous Irishman. Our principal entertainer was Miss Helen Murphy, a girl of much good sense...they had been two years in the country and possess a fine farm."

At this time in California land was of low cost. Weber bought the eight thousand seven hundred-acre *Rancho Canada de San Felipe y Las Animas* near the *Ojo de Coche* for four hundred dollars (it became better know as the Weber or the Coyote Ranch). With the help of his brother-in-law, Dan, who was working the *Llagas Rancho*, it was used principally for grazing and horse breeding. More important to Charles was its proximity to the elder Martin's *rancho* which provided opportunities for him to visit Helen.

In June of 1846, William Gulnac, Weber's business partner, despaired of finding settlers for the *Franceses rancho*. Arriving emigrants were reluctant to settle there when they learned of the hostile Indians in the area and that one of his settlers was murdered by an Indian who then set his shanty on fire. Consequently, he sold his share to Weber for two hundred dollars. Now full owner and ready to go, Weber's first action, taking a page from Captain Sutter's book, was to meet and make friends with the much-feared José Jesus, chief of the Miwok and successor to the famous Chief Estanislao (the Stanislaus River and county were named for him).

José Jesus was over six feet tall, well built, proud and dignified in manner. Although, like Estanislao he hated the *Californios*, he usually dressed in the colorful attire of a Spanish grandee. He and Weber became good friends and, no doubt influenced by Weber's lavish gifts, José Jesus signed a peace treaty with the white man that was never broken.

The trip to the altar Charles and Helen planned for so long sometimes seemed like an obstacle course. There was the Mexican War in 1846. Because of his defection to the Americans, General Castro considered Charles, a naturalized Mexican citizen, a traitor to California. When Castro and Alvarado, after surrendering Los Angeles, left for Mexico, they took Weber as far as the Mexican border. Little did Helen, who was already concerned for his safety, know that he was left to find his way back without food or water. He arrived at the *Pueblo de Los Angeles*, debilitated and ill.

On his return to the *Pueblo de San José*, Charles was appointed captain of the San José Militia with Helen's brother, John, his lieutenant. This was not a military operation but rather its purpose was surveillance. They feared reprisals from the *Californios* because the *Californios* resented the treatment they had received from the Americans. Against the urging of Charles and members of Helen's family, she and her father chose to stay at the *Ojo de Coche* for the duration of the war, even though other foreigners moved to the *pueblo*.

The following year, with the Mexican War behind him, Charles decided that before he could settle down to marriage he must first start developing his city (which he named Tuleburg because of his use of tules in building corrals for his cattle and horses and for building cabins). Located on a succession of sloughs with backwaters formed by the junction of the San Joaquin and Sacramento Rivers, its harbor was a bay four miles long and three hundred yards wide.

Charles's first step was to have the land surveyed and a plan laid out for the proposed town. Its name was changed to Stockton, a decision he would regret. During the war, after his release by Castro, he had met in Los Angeles Commodore Robert Field Stockton who sounded interested in Weber's plans for a town. The commodore said he would send him a ship to run passengers between Tuleburg and San Francisco and that he would help him to get a patent on his land—both of which turned out to be just so much talk.

To encourage settlers, Charles offered emigrants free land. Yet he met the same problem as Gulnac. The prospective settlers feared the unfriendly Indians of that area. When he made friends with José Jesus, he explained to him that the Americans wanted to be his allies and friends. He had no trouble with the Miwok, José Jesus's tribe, but remembered Sutter's advice when Indians, mostly the Chowchilla and the Polo tribes, stole cattle and horses or raided their huts. Charles and a group from neighboring *ranchos* including Captain Sutter, Robert Livermore and John Marsh would band together and attack them. They were thus prevented from having feasts on horsemeat, which they dearly loved.

On his trips to see Helen at the Ojo de Coche, Charles would give her progress reports. She approved of his plan of attracting settlers by giving them lots in town and acreage in the country as well as sometimes even giving them agricultural equipment and horses.

Yet it was this excessive generosity that would bother his brother Adolph when he came to visit in 1853. In a letter to his brother Phillip in Germany he wrote, "People know him for his loans to unworthy persons who suck the very marrow from his bones." He went on to write that Charles was never left alone—constantly hounded by someone wanting or taking something for nothing.

The discovery of gold, only fifty miles from Stockton at Sutter's Mill in January of 1848, proved to be the turning point in the fortunes of Captain Weber. First he formed the Stockton Mining Company in which his future brothers-in-law, John and Dan Murphy, were stockholders. Having gotten a jump on the masses of prospectors who were to come, they did well. But Weber soon decided the big money would be in supplying the miners.

Not only did Weber go into merchandising, but his friend, José Jesus, provided Indians to dig gold for him. They did their job well and as a result he became an extremely wealthy man. At the same time, his town of Stockton was to experience a population explosion. With its natural harbor, Weber had chosen his location wisely. It became the point of debarkation for miners who came up the river from San Francisco by the thousands. He couldn't build houses or tents fast enough to keep up with the demand. By late 1849 the population had grown to five thousand. Recognizing the need for means to transport supplies and merchandise as well as communication with San Francisco, he bought a two-masted, Oregon-made ship, the *Maria*, which became the first regular mail packet to sail between the two towns. By this time the skyline of the harbor was outlined by the tall masts of barques, brigs and schooners. But with the trip from San Francisco sometimes taking ten or more days, steamers would replace the sailing ships in 1850.

That same year Charles built for his bride-to-be a home of *adobe* and redwood on Weber Point between two arms of the slough. The bricks for the chimney were ordered from Plymouth, Massachusetts at an exorbitant price. It was said to be one of the great houses of the day.

In October 1848 the couple's long-running romance finally culminated in marriage. A joyous occasion—Helen was surrounded by her loving family in the *adobe* church of St. Joseph where Father John Nobili, later founder of Santa Clara College, performed the ceremony. Charles, who had earlier in the year joined the Catholic Church, was well liked by the Murphys. Her brothers, Martin, Daniel and Bernard, and her brother-in-law, Thomas Kell, signed as witnesses. Her misty-eyed father was torn between joy for his daughter's happiness and sadness because he knew how much he would miss her. Distance, however, would be no problem to this doughty Irishman, for he would ride his pony to Stockton for visits.

The home Charles Weber built for his bride in 1851 at Weber Point was one of the finest homes in California at the time. COURTESY HELEN WEBER KENNEDY COLLECTION.

Helen was pleased with the way her husband had planned their home, with its separate kitchen and dining room connected to the house by a covered passageway—a reflection of his years in New Orleans. Charles spared no cost. The mirrors came from France, the piano from England, china, crystal and beautifully crafted furniture from Europe—all in superb taste. From the cupola in the evenings they looked out on the ships coming up the channel and enjoyed the view of Mount Diablo. Yet with all this she missed her close-knit family.

On their first visit she proudly showed off her beautiful home. Afterward, Charles entertained them at the Stockton House, later to become famous as the St. Charles Hotel. A local newspaper reported:

The first entertainment that was respectable in character... the guests were received by E.M. Howison and H. Tabor Boorman.

Among those listed attending the dinner from San Jose were Mary and Daniel Murphy, Virginia and John Murphy and a Miss Wilson. "At two o'clock in the morning all sat down to a superb supper and then finished the night in dancing."

With horticulture a consuming interest of Charles, one that Helen shared, he made their home on Weber's Point a showplace. When Bayard Taylor, correspondent for Horace Greeley's *New York Tribune*, visited them, he delighted in seeing how the land between the two sloughs was now transformed into a large garden.

"There is no more beautiful villa in existence," he wrote, describing also a thick hedge backed by a row of semi-tropical trees that surrounded the peninsula—in addition to roses, lemon verbena, heliotrope, and fuchsia growing against the walls.

At the same time, Charles was developing his city by giving land for schools, municipal buildings and the San Joaquin County Fairgrounds. He and Helen had their cultural interests also. They served them by bringing music and theater to Stockton. Until the year after Helen arrived there was no Catholic Church. That was remedied when a French Jesuit, Father Blaive, visited Stockton with a group of emigrants. Captain Weber not only gave him land, but also built for him a church in a square, which he surrounded by shade trees. Here Helen and Charles attended Mass. However, Charles did not restrict his generosity to the Catholics. He gave land sites to the Congregational, Lutheran, African Methodist Episcopal and African Baptist congregations.

In 1853, Charles's father, the Reverend Karl Weber, who had received no word from him since he left the family home at Zweibrucken fifteen years earlier, asked his son Adolph to make the trip to California to learn for certain if reports he had received that Charles was alive were true. Accompanying him was his cousin, Julius Dauber. After greeting them warmly, Charles introduced the young men to Helen and their two children, Charles, Jr. and Julia (a third child, Thomas

Jefferson Weber arrived two years later} and showed them around Stockton.

Adolph wrote to his father about Charles's family, especially about Helen, with whom he had become close friends. He told of her efforts to get Charles to write to them—of her desire that they visit Germany. As a member of a large, close-knit family, he said she wanted to have closer ties with the Webers. Adolph had many lengthy conversations with his brother, yet he was unable to cast any light on why Charles had never written to his parents.

Finally, Adolph decided it was time to present himself to Helen's father and other members of the Murphy family—a new experience for this German native. After hunting and fishing with her brother, John, he continued on to the San Martín ranch of her brother, Dan. There he had a liberal introduction to life on a California cattle ranch, one he found fascinating.

As the '50s moved into the '60s, Charles was beset with problems that seemed to pile one on top of another. There were boundary disputes, complaints from residents, and political problems, but most of all, squatters. Many Americans came expecting to get rich quick and when that didn't happen, assumed that all land belonged to the United States and moved in. On one occasion Weber planned to improve a road between Stockton and French Camp but was prohibited from crossing his own land by a squatter with a gun. He finally had to use costly legal means to evict them.

Compounding his problems were the delays in his efforts to obtain a U.S. patent on the *Franceses* land grant without which he would lose everything. After his claim to the *Franceses* was confirmed by the U. S. Board of Commissioners, the U. S. Court of Appeals challenged it. With a battery of lawyers, he finally received U. S. Supreme Court confirmation in 1855. Helen breathed a huge sigh of relief. She felt deeply his frustrations and the lack of appreciation for his intensive efforts to develop Stockton.

The years of 1861 and 1862 brought heavy floods. The water stood eighteen to twenty inches deep in the Weber house. This seemed to Helen to be the right time to move to the *Las Animas Rancho* where Charles would be away from the trials of Stockton and could take things easier. Also, Helen would be closer to friends and family. But there was no way Charles would consider leaving Stockton.

Helen began to spend more and more time at the summer house in the Santa Cruz Mountains that Charles had restored for her and the children on land he had purchased from a squatter on Dan Murphy's Ranch. She sensed that he was undergoing a personality change—becoming increasingly withdrawn and caustic to her. The knowledge that his mother had spent years in a mental hospital gave her cause for concern, and rightly so. Charles eventually became a recluse, stopping communication with his family and friends. Helen worked doubly hard to keep a pleasant relationship between him and the children as well as keeping his business affairs in order.

Death came suddenly to Captain Charles Maria Weber in Stockton in February of 1881. At his funeral in St. Mary's Church, he was eulogized for his vision, his dedication to the building of Stockton and his unparalleled benefactions. Archbishop Alemany spoke with feeling in his sermon as he said, "He was a noble-hearted man and well performed his part as a good citizen and as a Christian." His eulogy, printed in the *Stockton Weekly Independent*, commended Charles for his care in nurturing his city of Stockton, for his honor and his valor.

In Charles's memory flags flew at half-staff throughout the city, courts adjourned and business houses closed from one to three in the afternoon. In the funeral cortege, led by the Stockton Band and marching to the Catholic cemetery, were men and women from all walks of life. They included public officials, fraternal organization members, and two hundred Chinese, most of whom had worked for him.

Helen survived her husband by nine years. Paying honor to Stockton's first lady on her death, flags of the city again flew at half-staff, as did those on the boats in the bay. A crowd estimated at one thousand people gathered outside the Weber home as the procession formed for the march to St. Mary's Church to remember the much-loved lady. One eulogist said that her life had been a labor of charity and that the poor would miss her.

The *Stockton Evening Record* reported:

A pioneer of pioneers. A woman of personal simplicity in an unostentatious association, she displayed a mental greatness unmistakable of a woman admired by all who knew her. She loved to witness the march of progress and felt a true pride in the city that was reared on the soil in which she had assisted in sowing the first seeds of cultivation.

CHAPTER TWENTY
The Martins and the Murphys

After the Stephens-Murphy party men came down from the mountain and assembled at Sutter's Fort, James (called Jim by the family), the patriarch's second son, reluctantly joined Sutter's army with the others. He had no choice. The alternative was to retrace his steps back to Council Bluffs. This was not a happy prospect for his wife, Ann Martin, who was eight months pregnant and already weary from the long trek across the prairie, much of which she walked.

As it turned out, Jim and his brother Bernard were the only members of the family to go all the way to the *Pueblo de Los Angeles* with Sutter. Meanwhile the elder and younger Martins and Captain Stephens, concerned about the women and children in the camp on the Yuba River, had met with Governor Micheltorena and received his permission to return to the cabin to rescue them.

Jim returned to Sutter's Fort after the signing of the Treaty of Cahuenga and found Ann had delivered a son. They named him Martin for his grandfather—one of numerous Murphy offspring to bear the patriarch's name. He now faced the question of where to go from here? He discussed it with John Bidwell, one of Captain Sutter's trusted men, who knew the country north of Yerba Buena. After Sutter purchased Fort Ross from Russia, he had arranged for the takeover. He told Jim that his sister, Mary, and her husband, James Miller, had gone with their children to San Rafael where there was timber.

That was all Jim, who had worked as a lumberman in Maine and Missouri, needed to hear. Accompanied by Ann, his four-year-old daughter, Mary Frances, and infant son, he set his course for San Rafael. After getting settled on the *Corte de Madera del Presidio* land grant of Dublin-born John Reed, who had preceded Timothy Murphy as the major-domo of Mission San

Rafael, Jim lost no time getting involved working in lumber.

For a time he and his brother-in-law, James Miller, worked together before the latter went into farming. In his *New Helvetia Diary*, Sutter noted that James Murphy and James Miller arrived from San Rafael to saw lumber. In May of 1847 he wrote of James Miller arriving from Sonoma with John Reed. Another entry indicated that Sutter sent Indians to Dennis Martin, Jim's brother-in-law who was lumbering on the San Francisco peninsula at what was then known as Greersburg, later renamed Woodside.

James Murphy was the second son of Martin, Sr. He shared ownership of Las Llagas Rancho with his brother, Daniel. FROM *PEN PICTURES FROM THE GARDEN OF THE WORLD* EDITED BY H.S. FOOTE, 1888.

Jim supplied the lumber for Yerba Buena's first wharf, the Leidesdorff Pier, and for the houses that were springing up like mushrooms in rapidly growing Yerba Buena. In Sutter's diary, he recorded that James Murphy and William Leidesdorff arrived with a half dozen others on the first steamship to arrive at New Helvetia from Yerba Buena.

Although Jim and Ann were happily settled on John Reed's *Corte Madera rancho*, the lumber boom was about to end. Gold frenzy seized San Francisco in 1848. His workers couldn't resist the lure of gold and took off for the Mother Lode. With an "if you can't beat 'em, join 'em" philosophy, Jim decided to take a fling at mining himself. He started off at Sutter's Mill, then moved to Hangtown, where he visited all the diggings in that vicinity. But he soon figured out that a man with a family needed something more secure. He stopped to see his older brother, Martin, at his *Ernesto Rancho* on the Cosumnes, and, observing how well he was doing selling cattle and wheat to the miners, decided to have a try at that, but not before he bought two leagues of the *Los Cazadores Rancho* adjoining Martin and Mary Bolger's *Rancho Ernesto*.

From there he rode down to the *Ojo de Coche* to see his father and to find out about land below the *pueblo*. Around the dinner table that night he talked of his predicament and said that he was thinking of raising cattle to sell in the gold country. His brother, Dan, just back from there and getting established in ranching, said, "Jim, I hear that Guillermo Castro is going to sell his father's San Francisco de Las Llagas rancho. It has twenty two thousand acres and connects with the Ojo de Coche on the south. Why don't you and father buy it together?"

Without hesitating, Jim said, "Sounds good. Let's go have a look at it."

They did and, without further ado, had a deal. The sale was witnessed by his soon-to-be brother-in-law, Charles Weber, and former San José *alcalde*, Charles White. The Murphy clan's *ranchos* south of San José now covered an area extending ten miles from the southeast towards San José and seven miles across the valley.

Jim returned to the Corte Madera and broke the news to a dubious Ann, who wasn't too sure she wanted to move to such a remote area away from her father and brothers. Nevertheless, they packed up their belongings and moved bag and baggage, including the latest family addition named Helen, to the Llagas where they settled in the Castro *adobe* on Llagas Creek.

Apparently life on a cattle ranch didn't appeal to Ann because when the five hundred-acre lots northeast of San Jose, made up of pueblo lands, became available, Jim bought five of them. On one he built a house and moved his family into town. There he planted one of the first olive orchards in Santa Clara County. With the Murphy penchant for farming, his ground was carefully and intelligently tilled. His neighbor on the adjoining land was Moses Schallenberger, the same young man Ann's brother, Dennis Martin, had tramped with through the snow to rescue at Donner Lake.

Dennis, meanwhile, acquired a portion of the *Rancho Canada de Raymundo* in an exchange with Juan Copinger for hardware he happened to be selling at the time, and later bought a piece of the Rancho Corte de Madera—both parcels of land located on the San Francisco peninsula. He became the first and most successful of the lumbermen in this area later known as Woodside. He not only built a house on San Francisquito Creek, but two mills, houses for his workmen and a large barn for stabling his oxen used in hauling lumber.

A deeply religious man, he then built a chapel in which he put expensive candlesticks and a crucifix. He enclosed the churchyard with a fence and a short distance up the hill he fenced in a cemetery. His was the first house of worship to be constructed between Mission Santa Clara and Mission Dolores. Archbishop Alemany, in dedicating the new building, named it St. Denis (spelled with one "n") in honor of its donor's patron saint. Dennis's father, Patrick Martin, was the first man to be buried in the little cemetery and Dennis the last in 1890.

A favorite worker on the ranch was Popka, orphaned son of Chief Licia of the Feather River Indians. Dennis had brought him at the age of ten to live at his ranch. He had him baptized James Martin in the little St. Denis Church. In a 1941 *San Jose News* story, Mrs. Fremont Older wrote of a visit she had with Indian Jim, as he was usually called, at an infirmary in San José when he was eighty years old. She said that he considered himself an Irishman, and a Roman Catholic, even to the point of celebrating St. Patrick's Day during his wild youth by getting drunk.

In her interview she wrote, "He sat up in bed and talked of his 'father,' Dennis Martin, and of his pride in knowing his catechism, and in being a good Catholic.

He showed me his crucifix and the broken rosary around his neck under his gray undershirt. He told me how good Dennis Martin had been to him, how he knew his alphabet and told me how to spell "cat, dog and sky." Unfortunately, Dennis wasn't able to help Indian Jim very much after he became the victim of a boundary dispute and lost everything including his improvements and a producing young orchard.

In 1872, because Jim and Ann's family had increased to seven children, they built a spacious home called Ringwood Farm at a cost of forty thousand dollars, and surrounded it with beautiful grounds so that it was one of the showplaces of the valley. Jim and Ann were unhappy when their eldest daughter, Mary Frances, who was born in Frampton and had crossed the plains with them, broke the news that she wanted to marry young Bernard Machado, a *vaquero* on the *Llagas rancho*. They apparently considered his education, or lack thereof, and social position beneath them. Mary Frances and her sisters had been good students at Notre Dame Academy. Jim's objections seemed a bit of a paradox considering that he signed his own name with an "x."

Ringwood Farm was the showplace residence of James and Ann Martin Murphy. FROM PEN PICTURES FROM THE GARDEN OF THE WORLD EDITED BY H.S. FOOTE, 1888.

Nevertheless, the two were married and established a home in Gilroy, where they raised a family. Eventually, with financial success, Machado became motivated toward education—served several terms on the school board and gave a tract of land for a school named for him.

No objections were heard from Jim and Ann when, in May of 1879, their youngest living daughter, Julia,

married Santa Clara College graduate Richard Fox. Fox was a native of Dublin, Ireland who, after losing his parents at the age of twelve was sent to California to be adopted by his uncle, pioneer horticulturist Bernard S. Fox. The elder Fox had been brought to California by Commander Robert Field Stockton to plant his San Jose gardens. Richard learned about horticulture from his uncle, and about the operation of his business. After his graduation from college he became his uncle's assistant in the management of his Santa Clara Valley Nurseries and Botanical Gardens, and upon his uncle's death, inherited the business.

As Santa Clara County grew, so proportionately did the demand for trees and plants, resulting in the birth of the California Nursery Company of which young Richard was principal stockholder and officer. The nursery property covered five hundred acres. The main building, surrounded by orchards planted by his uncle, stood at the far end of an avenue of stately evergreens. Across the road stood the botanical gardens, a showplace of tropical shrubs and flowers. On the other side of Coyote Creek another tract was converted into orchards and small fruit farms.

Although Julia inherited a parcel of the Murphy family property, she was a resolute woman, never known to change her mind. Apparently she had a grievance against her Murphy parents, because she never spoke of them to her children. Likewise, she had very little contact with her brothers and sisters although on record is a deed in which her youngest brother, Daniel J. Murphy, who inherited the Milpitas property, granted her a parcel of land along the San Jose and Milpitas Road. Young Dan inherited the Irish love for horses, and on his Milpitas land developed the Moorland Stock Farm in 1893.

Julia's sister, Lizzie, married Henry Bull (half brother of Mary Fisher Murphy (Mrs. Daniel) son of Liberata Fisher Bull and her second husband, Henry Bull). Their daughter, Anita Bull Fife, would come to the house for occasional visits, usually with her cousin, Willie Murphy Bassett, the only child of James T. Murphy, youngest son of the younger Martin. Anita's son, Jack was a mining engineer working in the Philippines at the time of World War II. During the Japanese invasion he was forced to join U.S. soldiers in the Bataan Death March.

Although Jim's estate at the time of his death in 1878, estimated to be three hundred thousand dollars, did not match that of his brothers, the younger Martin and Dan, it was nothing to be scoffed at. The beautiful residence drew its last breath in the earthquake of 1906.

Moorland Stock Farm of D. J. Murphy, son of James and Ann, was located in Milpitas, Santa Clara County. COURTESY MARJORIE PIERCE.

Also destroyed in that famous quake was the home of Julia and Richard Fox. They took over the Chinese cookhouse on another family ranch on Coyote Creek on a temporary basis until they could build a new house. But with her good taste and ingenuity, Julia managed to create a fine home so that they never got around to building another.

Julia and Richard's eldest son, Bernard Simon Fox, inherited the Murphy love of adventure. Upon his graduation from Santa Clara College in 1897, he took off for Mexico, near Chihuahua, before going into the mercantile business in Casas Grandes, Mexico. He left that for more exciting employment as scout and interpreter for General John Pershing during the

Pancho Villa uprising of 1916-17. President Wilson, disturbed over Villa's execution of sixteen Americans, sent an expedition of ten thousand soldiers to Mexico, headed by Pershing, to capture the renegade Villa. The mission proved unsuccessful because of Villa's political ties. Because of the peasants' loyalty to him they would never reveal his whereabouts. It was said if Pershing had taken one hundred Texas Rangers down to Chihuahua, he could have succeeded. Bernard returned to the Hearst Ranch to become its manager. In 1926 he moved to Texas where he served as a judge for sixteen years.

His sister, Ada, the irrepressible eldest Fox daughter (who marched to a different drummer from the rest of the rather dignified family) also moved to Mexico with her husband, Carl Schilling. There, Carl, whose father had a gun store in San Jose, taught her how to shoot. The diminutive Ada, who barely reached four foot eleven inches in height, became such a crack shot that the couple formed a vaudeville act. Dressed as an Indian maiden in a fringed leather skirt, Ada shot a cigarette out of the mouth of Carl, who was dressed in western attire with fur chaps. They played the Keith-Orpheum vaudeville circuit for over ten years opening U. S. Senator James Phelan's posh Victory Theater in San Jose in 1899 and playing the Panama-Pacific International Exposition in San Francisco in 1915.

Ada delighted in nature, especially animals and birds, and after she inherited the family home she had it full of them. In an aviary outside she housed parakeets, canaries, and finches while peacocks and pheasants roamed the grounds. In an indoor cage lived a parrot which would call out to Ada, "Hello, Love." Ada did it her way and got away with it. The family attended Sunday Mass at St. Patrick's Church and, inasmuch as she was always late, she double-parked—usually next to the second or third car from the corner. There the car would stay while the family congregated after the service to visit.

One time, Ada's niece, Ora Woodward, remembers going to downtown San José with Ada. There were no parking places near the bank, so she blithely parked in front of a fireplug, walked over to the policeman on duty and in her bright friendly manner said, "Watch my car for a few minutes. I won't be long." Then, looking at his name badge, she said to the befuddled young officer, "I knew your grandfather well." When they came out of the bank she thanked the officer and drove off.

CHAPTER TWENTY-ONE
The Soaring '60s and '70s

By 1860, the Murphys were all solidly established and still moving right along—doubly happy over the arrival of steamers at Yerba Buena and the Pony Express service from St. Louis to San Francisco that revolutionized communication. Four years later the San Francisco & San Jose Railroad was completed. Even better news, President Lincoln signed the Pacific Railroad Act in 1862, making possible the transcontinental railroad to be built by the Union and Central Pacific line (later Southern Pacific) which would facilitate the younger Martin's shipments of grain and cattle. Because this line would end at Sacramento, Martin's good friend, Peter Donahue, built his Western Pacific Railroad from San Jose to connect in Sacramento with the Central Pacific.

The patriarch, by this time, was living with Dan and Mary Fisher and their three children at their home in San Martín on the *Llagas rancho*. He kept occupied tending to his business affairs, including his cattle and steam-powered mill in the redwoods. Much of his time was spent visiting the families of his daughters and sons. The welcome mat was always out when he arrived at Martin and Mary Bolger's Bay View Ranch. Though his face was now lined and leathered from the years in the sun, his smile was warm. He accepted good-naturedly the family's teasing about the demijohn he carried on his saddle—a family joke, because they knew the patriarch never touched liquor.

The younger Murphys' eight lively children grew up in this house where hospitality was unlimited—and everyone enjoyed a party. In Alfred Doten's Journal he tells of the Murphy boys coming to get him to play fiddle for a party. Besides spirited singing, the ladies were kept up dancing cotillions, reels, waltzes and polkas. According to Doten, "Irish whisky was plenty and all went in freely for having a good time. It was a real Irish scrape."

On his deathbed the Murphy's good friend, Truckee, asked that his granddaughters be taken to the sisters' school in San Jose. After only two weeks, the girls were asked to leave because influential parents didn't want their daughters going to school with Indians. One of his granddaughters, Sarah Winnemucca, would become famous for her defending of the Paiute tribe and be invited to the White House to meet the president. COURTESY NEVADA HISTORICAL SOCIETY.

Made welcome were rich or poor traveling along El Camino between San Jose and San Francisco. The younger Martin believed in helping a man down on his luck, but in consideration for his self-esteem, would offer him some task. His work ethic, according to his son, Barney, in his Bancroft Library dictation, was that a person should earn what he gets.

In this same dictation Barney described his mother. Noted for her great heart, she used to say, "The only reason the Lord gives you anything is so you can share it or give it to other people in the best possible way."

One day the usually even-tempered Mary Bolger became angry when she learned that Sarah Winnemucca and her sister, granddaughters of the Murphy-Stephens Party's good friend, Chief Truckee, had been requested to leave Notre Dame Academy only two weeks after their arrival. On his death bed the chief had asked a friend named Hiram Daniel Scott (Scotts Valley) to take his granddaughters to the sisters' school in San Jose. Truckee had seen the beautiful school when he made winter visits with his family to the home of Peter Quivey, whom he had led across the Sierra in 1846. On those occasions, in native fashion, Truckee would set his wickiup in the Quivey's backyard.

Mary Bolger remembered well, when the Murphy-Stephens party was at Humboldt Sink without a clue as to which way to go from there, that it was Truckee who directed them to a pass over the Sierra. She lost no time in paying the sisters a call. What upset her most, she told them, was their explanation that some of the wealthy parents of students had threatened to take their daughters out of school because they didn't want their children associating with Indians. Needless to say, the sisters, who remembered the support of the Murphys in establishing their school and found Mary Bolger's daughters, Elizabeth Yuba, Mary Ann and Nellie to be model students, regretted having yielded to pressure from some of the students' influential parents. Their embarrassment only increased when Sarah became one of the outstanding leaders of the West.

Martin Murphy, Jr. built the first brick building in San Jose, which housed the county courthouse. FROM *OLD SANTA CLARA*

Chief Winnemucca, father of Sarah, succeeded his father Truckee as chief of the Paiutes. COURTESY NEVADA HISTORICAL SOCIETY.

Although she adopted the white people's dress and culture, Sarah became a fearless champion of her Paiute tribe. She lectured everywhere on the cause of her people and on their treatment by the government. Washington became alarmed and invited her and her family to the White House. There, President Hayes signed an executive order allowing the Paiute to leave Yakima where they were interned and granting them land which they unfortunately never received. She later appeared before a congressional committee to speak for a bill that would grant the Paiutes land of their own. The bill passed but the Secretary of the Interior refused to implement it.

By the early '60s, the younger Martin's business acumen had made him a man of considerable wealth and standing in the community. He was the first to import Norman horses from his native County Wexford and the latest and most efficient in farm machinery from the East by way of the Isthmus of Panama.

His real estate ventures had moved into urban development. Instead of building with the traditional *adobe*, Martin was the first to use brick. About the same time a block of brick buildings he had built was completed, the newly organized city council's lease for office space ran out. He volunteered to refurbish the upstairs of his new building to be suitable for government offices, and give them a five-year lease at $190 per month.

Soon to follow were the Jefferson and Washington Blocks and the Murphy Building, in which the first California Supreme Court convened under American jurisdiction. In the devastating earthquake of 1868 its heavy cornice, an architectural feature of the building, crashed to the ground. With all his building activities still foremost with the younger Martin was agriculture which inspired him to build the City Market, an inner-city marketing center for the exchange of goods and services.

With all his successes, the younger Martin had other concerns. When California became part of the United States it was necessary for land grant owners to obtain patents to their property from the new government. That entailed, in Martin's case, fifteen years, which meant he and other land grant owners in similar circumstances couldn't sell off any property until they obtained title.

His second problem involved squatters moving on his land. According to the 1848 Treaty of Guadalupe Hidalgo, "property of every kind belonging to Mexicans already established there, shall be inviolably respected." That decree went down the drain in 1851 when Congress passed an act establishing the Federal Land Commission to "ascertain and settle private landward claims in California." Thus, with the stroke of a pen, Congress declared all the land titles to be suspect. This offered encouragement to squatters hoping to seize the title if a claim was rejected.

This was a hardship for the *Californios*, because the legalities entailed a long, costly process. Most couldn't speak English and didn't understand the new law which necessitated that they hire lawyers who would take parcels of land as payment—until finally many of the grantees lost it all. The younger Martin, who could afford high-powered lawyers, obtained patent to his portion of the *Pastoria de Las Borregas* (Bay View) in 1865, but Mariano Castro, owner of the other half of the land grant, didn't receive his until 1881.

With the appreciation of land values, Bay View was getting too valuable for grazing cattle. The younger Martin turned his eyes southward to buy the seventeen thousand acre *Rancho Santa Margarita* in San Luis Obispo County for forty-five thousand dollars from its grantee, Don Joaquin Estrada, who was having problems establishing boundaries.

These were hard times for the California ranchers. During the great drought of 1862-64 cattle died by the thousands and a strange malady killed horses in large numbers. Land values plummeted. The younger Martin was fluid financially, which enabled him to purchase the smaller four thousand-acre *Rancho Atascadero*. But in 1864 his really large acquisition was the thirty-nine thousand-acre *Asuncion* land grant from a man named Farrell for one thousand dollars. But he wasn't through in that area yet. He subsequently purchased the twelve thousand-acre *Cojo rancho* at Point Concepcion in Santa Barbara County. The acquisition of all this land, on which thousands of cattle would graze, totaled seventy-two thousand acres. Added to his Santa Clara County and other properties, it made him the largest landowner in California. And so it came about that the Murphys who left Ireland, where they were leasehold farmers, were now large landowners.

On a much smaller scale, but one that would eventually become extremely valuable to his heirs, was the younger Martin's purchase of a piece of the *Milpitas Rancho*. His eight hundred acres spread out from the hills north of Berreyessa Creek. James T. Murphy, youngest son of Martin and Mary Bolger, inherited this property. When he died in 1898 James left half his estate of one hundred fifty thousand dollars in trust to his only daughter, Mary Wilhemina Bassett, stipulating that if she left no heirs it would go to sixteen nieces and nephews. But at Wilhemina's death in 1968, the nieces and nephews he listed were all deceased. Consequently, by a court decree, the descendants of Martin, Jr. became the heirs. In the years since James T. Murphy's death the agricultural land has become valuable industrial property estimated at one million five hundred thousand dollars in liquid assets and two million dollars in property.

Murphy relatives, the Sinnotts, also benefited nicely from their purchase of Milpitas land. In 1856 after farming in Mountain View for a half dozen years, the younger Martin's cousin, John Sinnott and his wife, Elizabeth, who was Mary Bolger's twin sister, purchased eight hundred acres. Their large family became pillars of the community. John was instrumental in raising funds to build Milpitas' first Catholic Church. His efforts received strong support from a member of the local Baptist church. In appreciation, John asked the bishop to name the new church St. John the Baptist for his friend.

In 1864, Bay View, which had been a stage stop on El Camino Real, became a flag stop on the San Francisco-San Jose Railroad, giving the Murphy family easy access to the railroad. The Murphys had given a right of way to the railroad and deeded land for a depot. The little community that developed around it, called Murphy's Station, would eventually become the city of Sunnyvale. Later on, he built Lawrence Station at the south end of Bay View. As one of the perks, in addition to his family members being able to catch the train in their own backyard, the younger Martin received a railroad pass, enabling him to visit his far-flung land grant properties. The completion of the San Francisco-San Jose Railroad and the Central Pacific, combined with the decline in gold mining, brought a large number of Chinese into the Santa Clara Valley, providing the younger Martin and other large farmers with access to cheap labor. The procedure, if one wanted to hire Chinese laborers, was to write a letter to Governor Stanford at Mayfield. Apparently Martin heeded that advice because he had a dozen Chinese living in an outbuilding at Bay View.

In 1865, the dark hand of death struck the Murphy family twice. Mary and Martin's eldest living son, Martin, who was helping his father run his farming and business enterprises, was stricken at the age of twenty-nine and lived only a short time. Three months earlier the patriarch, Martin, Sr., had succumbed at the age of eighty in the home of his eldest daughter, Margaret Kell. Thus, he was spared the heartbreak of knowing the loss of this outstanding young man.

The death of the kindly, generous Martin Murphy, Sr., who, during his twenty years in California, became famous for his Irish hospitality, touched not only his family but the whole community. Courts adjourned for his funeral that took place in the *adobe* church of St. Joseph where, until six years before his death, he had ridden horseback the twenty miles from his ranch to attend Sunday Mass. The church was filled with mourners paying tribute to this valiant pioneer who, at the age of fifty-nine, had led his family across the plains to the Promised Land of California. His close friend, Father Peter de Vos, in his eulogy, described him as an Irish patriot, citizen of Canada, of Mexican California and finally, permanently, of the United States.

The *San Francisco Courier* wrote:

He was strictly upright and honorable in all his business transactions, generous, hospitable, and kind hearted. His house was always open to the friendless stranger, many of whom remember with gratitude his acts of disinterested kindness; energetic and enterprising a large property was the result of his labors in the evening of life.

The '70s brought further good times, but sad times as well for Martin and Mary Bolger. They again suffered the loss of a child. The beautiful, intelligent Elizabeth Yuba, her father's favorite, succumbed in 1875. She had separated from her husband, William Taaffe, and moved home with her four children from their *La Purisima Rancho* which had been a wedding gift from her parents. Her husband followed her in death only a few years later. Thirty years after their arrival in California Mary and Martin were again raising small children—the Taaffe boys, William and Martin, and their sisters, the twins, Mattie and Molly.

The Murphys had the Irish love for politics in their blood. The first meeting of the Democratic Party in the middle 1850s was held at Bay View, and the Murphys remained faithful to that party. Although the younger Martin never sought an office, he strongly supported the Democrats who were known to the *Californios* as the *Sabe Muchos* (Know Much) and the Republicans as the *Sabe Nadas* (Know Nothing).

In the late '70s a "no rent" agitation commenced in Ireland. A member of parliament came to San Jose to deliver a lecture. Barney Murphy, mayor at that time, was asked to preside. Concerned that his father, with his many tenants on rented land, might not be sympathetic to the party, he rode out to the ranch to talk to him about it.

The young Martin said to his son, "You go on and preside. When America gets like the landlords in Ireland, the quicker they are put out, the better."

There was no doubt that the Murphys had a feel for politics. Barney not only served four terms as mayor of San Jose, but in the assembly and the state senate. Governor Booth appointed him lieutenant colonel on his staff, Governor Pacheco appointed him colonel and Governor Irwin, judge advocate with the rank of colonel. His brother, Patrick, served three terms in the state legislature and again in the assembly, as well as brigadier general in the National Guard. The youngest Murphy son, James would eventually serve in the state senate for Santa Clara County. With his mother's Murphy blood in his veins William James Miller, who learned a lot about life at the age of twelve during the tortuous frozen winter on the Yuba, couldn't resist a fling at politics and was elected to serve in the state assembly representing Marin County in 1869. He and his cousins, serving in the assembly at the same time, enjoyed reunions, sharing memories of experiences crossing the plains and catching up on their respective families.

114

CHAPTER TWENTY-TWO

Civil War

With the outbreak of the Civil War, loyalties in San José were divided between Union and Confederate causes with strong feelings on both sides. The Union sympathizers had strong support from *San Jose Mercury* editor, J. J. Owen, an avowed abolitionist, who blasted the Confederates with his editorials. After the assassination of President Lincoln, he advocated the hanging of Jefferson Davis and General Lee. Southern supporters felt they were unjustly treated.

During the assassination furor, the elder Martin's granddaughter, Annie Fitzgerald, saved the day for the Notre Dame convent. The sisters, unaware of protocol, had failed to drape the convent doors with black. Gossip mongers spread the word that there was rejoicing within the convent walls over President Lincoln's death. As a result there were threats to "burn the convent down" and "drive out the traitors."

Soldiers trained on a street in San Jose as a precautionary measure should California be attacked from the Pacific Ocean. COURTESY CLYDE ARBUCKLE COLLECTION.

Concerned for the sisters' welfare, Levi Goodrich, the school's architect and neighbor, suggested that one of the students write a poem. Annie Fitzgerald, who would one day become one of the Notre Dame order's outstanding nuns, wrote an eloquent piece dedicated to President Lincoln. It was read at the end of San Jose's memorial ceremony by *Mercury* editor J.J. Owen. He was so moved by it that he ordered black bordered copies printed for everyone at the service.

The College of the Pacific had its problems, too. A schism developed over the flying of the United States flag. The Southerners, who made up half the student body, demanded that the flag be taken down. After the president of the college sided with the Northern contingent, it was necessary for a student to guard the flag day and night. At the commencement exercise a student wrapped himself in the flag, while the Southerners did everything they could to disrupt the proceedings.

Sister Anna Raphael was one of Notre Dame's outstanding sisters. She was a student at the school at the time Abraham Lincoln was assassinated and was asked to write a poem. The publisher of the San Jose Mercury, *J. J. Owen, was so impressed that he not only read it at Lincoln's memorial service but had copies made so every resident of San Jose could have one.* FROM *LIGHT IN THE VALLEY* BY MARY DOMINICA MCNAMEE, S.N.D.DEN., 1967.

At Santa Clara College calm prevailed even though the school's loyalties to the Union cause might have been contested. There existed, in a college safe, three thousand dollars in Confederate bonds though the safe held an equal amount in Federal greenbacks.

The closest the college came to becoming involved in the war was when one of its faculty members, Joseph Bixio, S.J., returned to Georgetown where he had been a teacher, to serve as chaplain, crossing enemy lines to attend to both the Union and Confederate armies. Suspected of being a spy, he was arraigned before Gen. William Tecumseh Sherman. Charges were dismissed.

Rumors circulated in the county of a plot by the abolitionists to burn the Missourians' Methodist Church South. An individual was hired to do the job, but he got his wires switched and set fire to the M.E. Church in downtown San Jose instead. Another case of mistaken identity concerned a Bishop Kavanaugh who came to California to attend the Southern Methodist conference and to ordain preachers. He was falsely accused of being a recruiting officer for the Confederates. After his arrest, when it became clear that he was innocent, he was allowed to come to San José on parole to preach.

According to Cora Baggerly (Mrs. Fremont) Older in a San Jose news story, a small amount of slavery existed in San Jose. "In those days servants were sometimes bought outright from captains of foreign vessels," she wrote. "A Peruvian named Juan was purchased by James Frazier Reed, the captain of a man-of-war in the San Francisco harbor. Juan worked with a Malayian [sic] as cook."

"In the early fifties Judge Kincaid, who lived between First and Second streets, had on his arrival from Missouri two Negro men, their wives and children. They were Judge Kincaid's slaves. "The judge subsequently found that he didn't need so many helpers, so he sold Abe and Sarah with their families to John Murphy." The Murphys lived nearby on Second Street.

"A Mrs. Ferguson brought from Kentucky a negro slave named Joe. Joe Ferguson was sold to James Frazier Reed for whom he worked for several years."

In the archives of the San Jose Historical Museum at Kelley Park, the Santa Clara County property rolls list six slaves and, in 1854-55, a person named Sampson Gleaves as belonging to a man named Findly.

As the war progressed, there was concern that the Confederates might attack from the sea. Military companies popped up here and there. Among the enlistees was John Murphy who was appointed captain of the Johnson Guard with a first and two second lieutenants. Even though the Murphys considered themselves Union Democrats, John's brother, the younger Martin, agreed with President Lincoln's proposal that compensation was due those who emancipated slaves.

Antonio María Pico was a San Jose delegate to California's First Constitutional Convention in 1849. FROM CLYDE ARBUCKLE'S HISTORY OF SAN JOSE BY CLYDE ARBUCKLE, 1985.

CHAPTER TWENTY-THREE

Barney the Politician

Although he started his professional life as an attorney, politics proved to be Barney Murphy's game. After passing the bar at age twenty-four, Martin and Mary Bolger Murphy's son, Bernard, entered into a partnership with a Santa Clara College classmate, Delfin Delmas. Although Delmas would become the country's most celebrated trial lawyer, Barney's legal career proved to be short-lived. A little more than a year later his brother, Martin, who managed most of his father's vast empire, died suddenly. His next oldest brother, Patrick, was managing the *Santa Margarita* and the *Atascadero ranchos* in San Luis Obispo, County, so it fell to Barney to take over managing his father's properties.

Bernard D. Murphy was known by his constituents as "Barney" and called "B.D." by the family. He served four terms as mayor of San Jose as well as several terms in the State Assembly and Senate. COURTESY RUTH MURPHY POLK COLLECTION.

Barney, a gregarious, outgoing young man, by this time had already decided that he really wasn't fond of law anyhow. In addition to his business responsibilities, he had an urge to get into politics, which proved to be the right direction for him. The Murphys were all active in the Democratic Party. Like so many Irishmen coming to the States, after lacking the right to vote in Ireland, they quite naturally gravitated to politics.

It wasn't long before Barney was in the thick of it. Unknown to him, he was nominated as the Democratic candidate for the state assembly. In his dictation to Bancroft Library, he said, "When I ran for the Assembly I was out at the ranch and didn't even know the convention was being held." By an overwhelming majority, he was elected to the Assembly of the State of California.

Moving forward fast—that same year he met and married at Mission Dolores in San Francisco the pretty, outgoing Annie Lucy McGeoghenan, who had come from New York with her family by way of the Isthmus of Panama in 1860.

In the assembly his legal knowledge was immediately recognized by his appointment to the judiciary committee. Representing his home territory well, Barney's first accomplishment was to secure the State Normal School for San Jose. By arguing with eloquence and conviction that San Jose was nearer the geographical and population center of the state and that the railroads were centered there, he overcame the opposition of other communities eager to win the prize. They included Napa County which offered one hundred thousand dollars if it were located there.

In 1880, the State Normal School burned to the ground. The legislators, hoping to seize the moment for their own special interests, introduced a bill to relocate the school. The battle of ten years before was renewed. With this turn of events, even though he was not a member of the legislature at the time, Barney dropped everything and headed for Sacramento to refight the old battle for San Jose. His mission was accomplished. The college stayed in San Jose and received an appropriation for the erection of new buildings.

Upon completion of his term in 1873 he and Annie Lucy moved to San Jose with their three children and built a spacious house on Third Street to accommodate his growing family that would eventually number nine children. Tastefully furnished, it became the center of social life in San Jose, and their hospitality renowned. Annie Lucy, a gracious hostess, shared her husband's conversational abilities, his caring for people and his concern for the less fortunate. These qualities, combined with her acts of charity, endeared her to his constituency.

Annie McGeoghenan Murphy, Bernard's wife, was a gracious hostess to the many official and social affairs associated with her husband's political duties. COURTESY RUTH MURPHY POLK COLLECTION.

In spite of the county's Republican 1870 majority of six hundred, Barney was elected to fill the unexpired term of Mayor Adolph Pfister. He obviously "owned" the town of San Jose that he had known since he was a child when it was still called the *Pueblo*, when its few houses were of *adobe*, when cattle roamed the streets, and oxen-pulled *carretas* were the principal means of transportation. He was subsequently elected by substantial majorities to serve four additional terms as mayor—all unsolicited.

In the late '80s, he attended the Democratic City Convention with a firm determination not to accept another nomination for mayor. He felt he had too many responsibilities, managing his family's estate, as chairman of the board of freeholders organized to form a new charter for San Jose and serving as judge advocate on the staff of the governor.

As reported in the account of the convention in one of the San José newspapers, he was standing at the rear of the hall when called upon for a speech. As he came forward he was enthusiastically greeted on all sides. With the usual twinkle in his eye, the talented storyteller told of a Missourian, carrying his pack, who came upon sixty men at a cross roads engaged in a fight. He asked, "Is this yere a free fight?"

They answered, "yes."

He said, "Consider me in." He sailed in knocking three or four men down, then was knocked down himself and before he could get up the crowd swayed backwards and forwards on top of him. After getting on his feet, shaking the dust from his clothes, he once again asked, "Is this yere a free fight?"

When somebody said, "Yes," he replied, "Count me out."

Murphy said he was not in the fight as a candidate, but as a member of the party. He exhorted them to place men on the ticket who would command the confidence of the people. The next speaker, Judge John H. Moore, commented that he agreed with everything Mr. Murphy had said except that he did not believe that the Missourian said, "He was out." He was inclined to believe that when a Missourian entered a fight, he went in to stay.

Ignoring Barney's firm statement to the contrary, his name was placed in nomination. After thanking his supporters, he gracefully declined, stating, "I am not able to discharge the duties of the office if elected. The first year I held the office it was with disgust to myself as I had not the time to give it the attention it justly deserved. My business has increased to such an extent that it now reaches from San Diego to the Siskiyou. In fact it is all outside business. It is absolutely necessary to give my attention to my private affairs, and I must therefore withdraw my name from before the convention."

His plea ignored, he was nominated by an overwhelming vote. In response he said, "I really don't know what to say. If I decline now it will be said that I am afraid. I think the best thing I can do is to be like Judge Moore's Missourian. He says a Missourian never considers himself out. "Gentlemen, consider me in." The meeting then arose and gave three loud and spontaneous cheers for Barney Murphy.

One of the first things he did on assuming office as mayor and in subsequent terms was to donate his salary to the Free Public Library that had been started by his predecessor. Oftentimes he supplemented it from his own purse. Losing no time in accomplishing his campaign promises, he graded and graveled the streets. He then had six sprinkling carts made and for the first time the streets were cleaned effectively. Another worthwhile improvement was the widening of the channel of the Guadalupe River, which had a habit of overflowing, causing disastrous results.

Many years later someone asked him if he was a reformer. He thought for a moment and then agreed that in some ways he was.

"At the time I was elected," he said, "the town was wide open. The hurdy-gurdies ran all day and all night. Women were employed in the saloons. I drove the women from the saloons and introduced an ordinance to close them at midnight."

He said, "the boys complained that I was unreasonable, but I pointed out that it was a fair division. It gave them half the night to get drunk and the other half to get sober." The ordinance passed.

In the late '70s and early '80s Barney's younger brother, James, owned the *San Jose Herald*. It wasn't paying. Barney knew they could have made the newspaper go if it attacked people, but it wasn't the Murphy style. He had an opportunity to have a little fun when a lumberman named Dougherty came by and asked him facetiously why he didn't get something lively in the paper. After he left, Barney went to the editor and told him to dig up something about Dougherty.

"I left town the next day, and just in time," he said, "for I learned Dougherty was hunting for me with a shotgun."

After his appointment by James Lick as trustee of the Lick Trust in 1870, he probably made his biggest impact on the city. Lick, the reclusive piano maker and grandson of a Hessian soldier who served at Valley Forge, came to California from Pennsylvania by way of Callao, Peru. With his nest egg of thirty thousand dollars acquired there, he made a fortune investing it in San Francisco sand hill lots. By the late 1870s, getting along in years, he began disbursing his money. He was especially interested in the building of an astronomical observatory that would have a telescope with the largest and most powerful lens known to science. Lick had first designated one of the peaks at Lake Tahoe for its site but fortunately kept an option to be able to locate it elsewhere.

Barney, eager to acquire this plum for San Jose, took Lick up to see the over four thousand foot high Mount Hamilton, located about fifteen miles east of San Jose. Lick was impressed with the magnificent view. Barney wisely pointed out it was near Lick's home and would be a fitting monument to his memory. This must have influenced Lick's decision because before his death in 1876 he asked to be interred in the observatory. Barney saw to it that his wish was granted. He was buried beneath the great refracting telescope.

Lick established a fund of seven hundred thousand dollars but with the stipulation that Santa Clara County would build a road leading up to it. The county was willing, but lacked funds in the treasury. Lick agreed to accept county warrants and loaned them the money. The legality of the arrangement concerned Barney because he was a trustee of the Lick Trust. But after due consideration and consultation with other legal minds, he decided the county would not repudiate the contract. Lick advanced sixty-five thousand dollars.

The Board of Supervisors was plagued with problems. The contractor failed and the county had to finish it with its own funds. A short time later, however, Barney was elected to the state senate and secured passage of a bill authorizing the board of supervisors to issue their bonds for the remaining thirty thousand dollars. Despite all its problems, Lick Observatory was a scientific achievement with the finest equipment attracting visitors from all over the world.

Lick Observatory on the summit of Mount Hamilton. As trustee for James Lick Bernard was able to secure the observatory that had the largest telescope in the world for San Jose.

A pet project of Barney's was the development of Alum Rock Park, a project which took twelve years to accomplish. When he first became interested, it was a dense, unexplored thicket, a habitat for birds and animals. Penitentia Creek ran down the base of the canyon. On its walls grew ferns and native shrubs and, in the spring, wild flowers. With the winter rains, the creek spilled over rocks and through narrow gorges sometimes creating waterfalls as high as one hundred feet.

The governor's board of commissioners appointed Barney chairman of the Alum Rock Park Commission with members General Henry Naglee, Edward McLaughlin, Dr. A.J. Spencer, Adolph Pfister and D.S. Payne. Their responsibility was to superintend the laying out of the park and the construction of a road leading to it. Recognizing the medicinal and health value of the many different kinds of mineral springs discovered within a mile of each other, Barney was responsible for the installation of free baths, modeled on those in Munich. When the park was finally ready, locals as well as residents of surrounding towns arrived by horse and buggy and by steam train. Still later they would come by electric trolley. They hiked the canyon,

splashed in its sulfur water plunge and baths, as well as enjoyed the band concerts, the merry-go-round, the camping grounds and picnic areas.

A man for all seasons, his captivating charm was equaled only by his tireless energy. In addition to the demands of his mayoralty and helping his father manage his vast estate, he became the first president of the new Commercial Savings Bank, of which his father was a stockholder. His great heart was never more evident than during this time. A friend said, "He gave more money than he spent. There was a sack of money behind the bank counter and any needy man or woman could ask and be sure of receiving help out of that sack." He had many friends among the Californios from his early days. During their hard times, Don Bernardo, as they called him, helped them out.

Another story told is how he loaded his pockets with coins before he left the bank to go to lunch at La Molle Restaurant. He knew there would be a line of needy people waiting for him to pass by in his long frock coat, string tie and boots. "Going along with his head bowed under his high crowned, broad-brimmed Stetson hat, he put a piece of silver to left and right into each extended palm and many a necessity depended upon

122

the daily walk of the man who knew so unostenta-tiously how to give."

Barney Murphy's contributions to his city and state are so numerous one wonders how one man could accomplish so much. He served on the staffs of Governors Booth, Pacheco and Irwin; was chairman of the board of Freeholders to frame a new charter for San Jose; and became presidential elector-at-large on the Democratic ticket in 1888. Barney stood by the San Jose Woolen Mills, when the company was on the verge of going under. In 1894, there was a clamor to make this pioneer the Democratic candidate for governor. A number of corporations and a group of politicians succeeded in defeating his nomination due to, among other things, anti-Catholic feeling of the time. With his usual grace, he insisted it was better that Jim Budd was nominated rather than he. "As a Catholic I might have been defeated although I had never been defeated for office before... Budd filled the bill much better than I, because he rode both horses. The Catholics looked upon him as a liberal and his father was an 'A.P.A.' (American Protective Association)." Budd was elected by a small majority.

From this point on it was all downhill for Barney. The Panic of 1894 brought on a depression that continued through most of the early '90s. During this time the Commercial Savings fell upon hard times. According to a great-granddaughter, his friend, James Henry, held a mortgage on the *Atascadero rancho*, but after telling Barney he would never foreclose, he did. He found himself a victim of friends, relatives and specifically the Shasta Lumber Company, to whom he had made unsecured loans. Yet, without ever a complaint, he paid to the depositors every dollar of the money he had stood good for by selling his holdings including his spacious San José home. With no place to live, his wife, Annie Lucy, who had charmed all with her social graces, took the younger children to the *Santa Margarita Rancho* to live with their uncle, General Patrick Murphy, while Barney looked for a job in San Francisco. The children took to ranch life, especially Helena, the youngest, who rode her horse to school. When the time came to rejoin their father, she put up a fuss. Ironically, this generous man who once had so much, now found himself with so little in material assets. Yet he held his head high in San Francisco and never a complaint came from him.

After standing staunchly at his side during the thirty-three years of their marriage, his beloved Annie Lucy died at the age of fifty of a strep throat. Deeply saddened, Barney carried on with the same dignity he displayed before losing his possessions. In San Francisco he held the position of Bank Commissioner. A raconteur, he amused his friends at the Palace Hotel telling stories—especially about the early days.

Shortly before his death in 1911, he said to his daughter, Evelyn Murphy, "I think I am getting to the end of El Camino Real."

Tributes to this dedicated public servant poured in to the newspapers:

a pillar of probity and popularity upon which the social and political fabric depended... he had the open, courteous, hospitable nature of the pioneers; he appreciated scholarship and was himself a reader of the world's best books.

For many years Barney seemed to be an indispensable part of the community.

He was our mayor whenever he wished to be; he was legislator whenever he wanted the office; he would have been governor without a dissenting vote if the County of Santa Clara could have made him so. Everybody loved and trusted Barney Murphy.

Editorials acclaimed the man and his career.

The Oakland Tribune:

He stripped himself of his fortune that no one might say that Barney Murphy caused him the loss of a dollar... It is a high tribute to a man to be able to say that he bore misfortune, poverty and ingratitude with the same dignity that distinguished him in wealth and power.

San Jose Mercury:

His qualities of head were of high order, and in themselves sufficient to warrant a high place in public opinion; but his nobilities of heart made him the object of common affection, and when fortune had gone and the good he had wrought as a public official and citizen had passed into forgetfulness, his acts of quiet kindness and open hospitality remained unforgotten, and the affection of his fellow citizens undiminished.

The honorary pallbearers included former classmates and friends at Santa Clara College; John Burnett, son of California's first governor, Peter H. Burnett; Thomas I. Bergin, president of the Santa Clara College Alumni Association; Delfin Delmas, Barney's first law partner, the country's most famous trial lawyer and a founder of Hastings School of Law. Family members included his only living brother, James T. Murphy, and the widow of his brother, John, Virginia Reed Murphy. His daughters present were Elizabeth (Mrs. Howard Derby}, Evelyn, Gertrude and Helena and his only living son, Martin.

CHAPTER TWENTY-FOUR

Pat Murphy and the Margarita

Patrick Washington Murphy couldn't have been happier when, in 1860, his father asked him to take over the management of his newly acquired *Rancho Santa Margarita* in San Luis Obispo County. It mattered not that it would mean his having to drop out of Santa Clara College or that this was an awesome assignment for a youth of twenty-two years. Pat, as family and friends called him, had many friends among the *Californios*. Having lived on a cattle ranch since coming to California as a child, he could compete with the *vaqueros* in lassoing a horse, and, as an expert horseman, could dash along at full speed, stoop over in his saddle and, with ease, pick up a rope, or any small object on the ground. The life of a *ranchero* would fulfill his fondest dreams.

General Patrick Washington Murphy inherited the Santa Margarita Rancho and became one of San Luis Obispo's best known citizens. An eloquent speaker, he served in both the State Assembly and Senate. COURTESY ANN DERBY JOY COLLECTION.

The younger Martin had purchased the more than seventeen thousand-acre *Rancho Santa Margarita*, from Don Joaquin Estrada, saving him from bankruptcy. The don had suffered from the heavy rains that year and from the costly lawyers' fees in trying to establish the land grant's boundaries while trying to obtain a U.S. patent. Compounding his financial problems was his lavish style of living—fellow ranchers from miles around gathered to enjoy his hospitality—the week-long *rodeos*, *fiestas*, horse races. On one occasion, he brought a circus from San Luis Obispo. Apparently Estrada believed the golden calf would never lose its luster. His brothers also acquired *ranchos*: Pedro, the *Asuncion*; Julian and Ramon Estrada the *Santa Rosa* and *San Simeon*, which William Randolph Hearst later acquired and on which he built his famous castle.

A short time later the younger Martin increased his land holdings in San Luis Obispo County to seventy-two thousand acres when he purchased the *Atascadero* and the *Asuncion ranchos* that lay adjacent to the *Santa Margarita*, *El Paso de Robles* and the nine thousand-acre *Cojo rancho* at Point Concepcion in Santa Barbara County.

Pat made his headquarters at the *Santa Margarita*, which had once been an *asistencia* to Mission San Luis Obispo de Tolosa. He lived in the house Don Joaquin had built called *Casa Estrada*. In the mission days the padres built a stone building containing a chapel, a section for granary storage, and housing for visiting padres who came to the *Margarita* to make their confessions. On the *rancho's* rich bottomland, which extended seven or eight miles along the Salinas River, they taught the Indians to plant corn and vegetables.

Travelers along the mission trail between San Luis Obispo and Monterey found the *rancho* a convenient stopping place. They later described the beauty of the Santa Margarita Valley, the mission Indians herding cattle and sheep over its rolling grazing land, the groves of oak trees, native shrubs and streams—all encircled by layer upon layer of hills and mountains as far as the eye could reach. As one writer, J. Ross Browne, expressed it in his book *Crusoe's Island*, "One of the earth's most treasured spots."

It was here, in 1846, after Commodore Sloat raised the American flag at Monterey, that General José Castro and Governor Pío Pico met on their way to the *Pueblo de Los Angeles*. They had hoped to avert a war with the United States. Pat's uncle, Dan Murphy, who served with John Fremont's California Battalion in the Mexican War, told him that during the battalion's stop at the *Margarita* on their way south, Colonel Fremont became suspicious when he learned of the Mexican officers' visit. He ordered the arrest of Don Joaquin Estrada, his major-domo, and his son-in-law, but soon after released them when they volunteered to join his California Battalion.

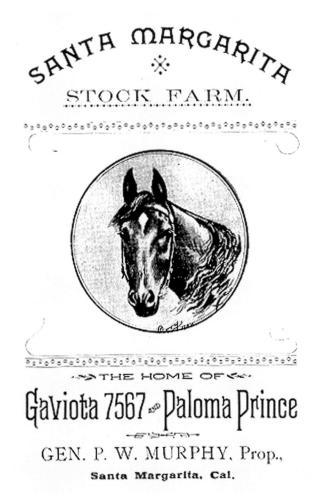

SANTA MARGARITA ⁕ STOCK FARM.

→THE HOME OF←

Gaviota 7567 and Paloma Prince

GEN. P. W. MURPHY, Prop.,

Santa Margarita, Cal.

A consummate horseman with the Irish love for horses General Murphy owned the Santa Margarita Stock Farm.

Even though the affairs of the *Margarita* started off smoothly, Patrick knew that the life of a *ranchero* had its ups and downs. Still, he had never experienced a rainstorm such as hit the valley shortly after he arrived. The torrential downpour continued day after day with the water racing down from the hills and canyons into the roaring Salinas River, which kept rising until it overflowed, eating away acres of land along its banks. *Adobe* walls and deserted mission buildings melted down to brown mounds. The *rancheros* lost thousands of cattle and sheep. Young Patrick was put to the test. He summoned his Indian *vaqueros*, and together they worked desperately to save as many of his stock as possible. Two years later the state was hit by a devastating drought. Starved for food, cattle and horses died by the thousands. Many *rancheros* lost their *ranchos*. That too passed and with sunny skies ahead he carried on his daily work.

The *rodeos* at the *Margarita* excited the whole community. Neighboring rancheros brought their cattle that had not been previously branded and strays that would be claimed by their respective owners. The bands of cattle were driven into the two hundred-foot by four hundred-foot corral on the bank of the creek that was surrounded by a heavy, ten-foot-tall fence. Pat and his major-domo directed the operation from the balcony of a two-story building in the shade of a centuries-old giant sycamore whose branches extended out fifty to seventy feet to form a canopy providing protection from the midday sun. Patrick branded three thousand calves a year—two thousand were his and the rest from neighboring *ranchos*.

After the branding came the fun part. True to Murphy tradition, Pat loved a party. His hundred or more guests dined on barbecued beef and danced the *jota*, the *jarabe* and *El Bolero Grande* to the music of guitars and the singing of the pretty señoritas. Yearling calves, cut in pieces, were barbecued over red hot-coals in a pit five feet long, four feet wide and three feet deep. A man on each side of the trench turned them continuously on long iron spikes until they were a rich brown. At the same time native California women were tossing *tortillas* and stirring frizzles and *chili con carne* in great *ollas*... all this and *aguardiente* aplenty.

While still young, Patrick Murphy became a man of distinction in San Luis Obispo. Governor William Irwin appointed him brigadier general in the Second Brigade of the National Guard even though he had no military training, prompting his American friends to call him "General." In 1865 he was first elected to the state senate representing San Luis Obispo, Ventura and Santa Barbara counties, and was, subsequently, elected to two more terms. In 1877, the voters of San Luis Obispo County elected him to represent them in the state assembly. He seemed to be involved in numerous enterprises, including being a stockholder and director of the new Bank of San Luis Obispo, and one of the incorporators of the city's water company. He spoke cultured Spanish with a slight Irish brogue, yet could speak the language of the *vaqueros* who affectionately addressed him as Don Patricio. With his polished manners, dark skin, bushy eyebrows, goatee, full mustache that curled below the corners of his mouth and his stylish tight-fitting clothes of the finest broadcloth and broad-brimmed black hat, he looked and acted the part of a Spanish grandee. Like an *hidalgo* of the old school, holding his hat before him, he kissed the ladies' hands, and ardently said he wished he could kiss their feet.

An unidentified San Luis Obispo native wrote, "This big, friendly, rollicking boy won the hearts of everyone he met—especially the Indians and the *Californios*." He spoke their language with the ease and eloquence of a native, patted the dark-eyed babies, complimented the tired mothers and kissed the daughters. "Patrick Washington," she continued, "arrived with a clatter of hooves in a cloud of dust. Every minute of his stay was an event to remember, and when he was gone, they talked of him for days on end. To them he was the 'Black Prince,' and his visits were milestones in the lives of both old and young in San Luis Obispo."

Calling him generous to a fault, the writer said that when he came to town the parasites would gather at the Cosmopolitan Hotel to greet him. Invariably he would say, "Gentlemen, you will drink a glass of wine with me." They did and usually had several—champagne, of course. The crafty bartenders never missed the opportunity of having empty champagne bottles on hand, which they charged to the chief. As befit his character, he paid for all in royal fashion. Inevitably there were vultures who hung around until he had several drinks before engaging him in a card game. After they had taken everything he had in his pockets, they would disappear.

The General's acts of charity were legend, especially to the poor and elderly. On one occasion, when he learned the work on the convent and school in San Luis Obispo was about to be suspended because of insufficient funds, he came to the rescue with his customary generosity. The building was completed.

Some locals lived off the ranch by killing his cattle and stealing from the granaries. One day his brother Barney's daughter, Elizabeth, was riding around the ranch with him. She said he stopped some *vaqueros* driving a cart of wood to tell them there was a barbecue down the road, saying, "You better go down and join them."

His niece noticed blood dripping under the wagon. When he returned she said, "You know perfectly well what was going on there. I could see the blood dripping down under the wagon just as well as you did. Why do you let them get away with it?"

He replied, "Oh, goodness, if you can't share what you have..." Besides, he told her, if he treated them kindly, it would be an insurance against their starting fires that could jeopardize the entire range.

Every year he invited all his nieces and nephews to spend the summer at the *rancho*. One time, when he had a big crowd there for a *fiesta*, one of his nephews remarked how nice it would be to go out on the bay in a boat. With that, Pat took the orchestra and the whole crowd, who were all dressed in their party clothes, to Santa Barbara and chartered a boat. The ocean suddenly became rough, causing many of the passengers to get seasick. Then it becalmed and they had to stay out there for a whole day. That was not one of the General's better *fiestas*.

With a natural instinct for politics, the peripatetic Patrick seemed to show up at every event. On election days he would arrive to electioneer for the Democratic ticket—knowing his influence especially with the Irish and the native Californians who invariably took his recommendations. Whenever his brother Barney was up for election Patrick was there.

The most important moment of his life occurred in 1870 when he took for his bride the beautiful Mary Catherine O'Brien, daughter of Dr. P.M. O'Brien of San Francisco, one of the founders of the Hibernia Bank. They had five wonderfully happy years during which Mary Kate, as he called her, endeared herself to all with her gentle manner and kindness—especially to the Indian *vaqueros* and their wives.

A source of great joy to Mary Kate and Patrick was to hear from her doctor that they were to become parents. One day, shortly before her baby was due, she handed Patrick a small bankbook of the Hibernia Savings and Loan Society.

"A long time ago," she said, "when the Hibernia Savings and Loan Society was just starting in San Francisco, I paid two dollars to become a member, and I put in three hundred." She showed him the entry in the bankbook—April 3, 1860. "I know it isn't much," she said, "but if anything happens to me I want our child to have it. Will you take care of it?"

"Of course," he said, and put it in a tin box that held the family records. He never thought about it again. It probably seemed like a trifling amount to a man who would inherit a cattle principality.

Mary Kate died in childbirth and was buried with her infant son in the Murphy family plot in the Santa Clara Mission Cemetery. Patrick was never the same man. Soon he started drinking to excess, though it seemed not to impede his accomplishments or diminish his charms.

The year following Mary Kate's death, he was invited to a banquet in New York. General McDowell, a stuffed shirt who had been in command of the Union Army, was there. Someone introduced Pat to him.

"General McDowell," he said, "this is General Murphy." After a peremptory handshake, the Union general said, "What division did you command in the Civil War?" Never at a loss for words, Patrick replied, "General I've made many a bull run, but I never ran at Bull Run."

In the mid-1880s the Southern Pacific was planning to run a railroad line just north of the Cuesta Grade and south of *El Paso de Robles*, which would pass through the *Santa Margarita*, the *Asuncion* and the *Atascadero ranchos*. They wanted Patrick to give them not only a right of way through the *Santa Margarita*, but six hundred forty acres for a town site. A compromise was reached by agreeing to give him a percentage of the financial rewards.

Benjamin Brooks, editor of the *San Luis Obispo Tribune*, a strong advocate for getting the railroad eventually to San Luis Obispo, wrote, "The town will be laid out upon a beautiful elevated plateau a short distance south of the old ranch house."

In a later *Tribune* story he wrote, "The rush will come for its choice business and residential sites, and we have the assurances of the owners of the Santa Margarita Ranch that a large part of that magnificent domain will be placed upon the market."

The following spring the Pacific Improvement Co. announced, "A Grand Auction Sale at the Santa Margarita"—stating that the Santa Margarita was the new terminus for the Coast Division of the Southern Pacific Railroad. A special excursion train was scheduled to leave San Francisco on Friday, April 19, picking up passengers at San Jose, Gilroy, Pajaro, Castroville, Salinas, King City and Paso Robles. The San Francisco firm of Briggs, Ferguson and Company would conduct the land sales. In the same ad General Patrick Murphy announced "a Grand Barbecue which he and his *vaqueros* would personally prepare."

The *Tribune* afterwards reported, "The biggest camp meeting ever held in the county. One hundred-sixty people came from the north by train. Visitors found the Santa Margarita Valley as crisp and lovely as old Dame Nature could dress it. The magnificent trees were clothed in their finest foliage, the grasses in their finest green and the flowers of varied hues and choicest colors.

Traveling the road over the Cuesta Grade from San Luis Obispo came a continual line of carriages, carts and wagons. Visitors gathered in the shade of oak trees or strolled around the town site and those interested had a great time selecting their favorite lot sites. As an added enticement the auctioneer announced that General Murphy soon planned to market ten thousand acres of choice farm land and that they anticipated it would make *Santa Margarita* a farming center.

When the smell of the barbecue drifted over to the auction site, the auctioneer wisely announced that the sale was postponed until after lunch. Benjamin Brooks wrote in the *Tribune* "If there is one thing General Murphy knows more about than another, it is managing an old-fashioned barbecue to a tee-y-te." Adding to the culinary delights the townswomen baked their favorite delicacies which they carried around to the tables in baskets.

Over one hundred lots were sold. It had taken the Southern Pacific two and one-half years to extend the railroad the short distance from Templeton to the *Santa Margarita*, and San Luis Obispo was still waiting for the train to come to them. At a meeting in San Luis Obispo in 1886 Pat gave a report on the subsidies given to the Southern Pacific Railroad while he was senator. He told them he had been in San Francisco talking with the railroad officials. As a result of that meeting, he warned them of the possibility that the Southern Pacific would bypass the town and build a depot closer to the coast if the town didn't offer S.P. some inducements. He recommended that they purchase acreage within the city for a station and that they build shops. C.H. Channing, a large developer of

California land, voiced his approval. Yet that time was still to come.

In June of 1890, San Luis Obispo sent a delegation to San Jose for a meeting at the Vendome Hotel to solve the financial problems encountered in continuing the railroad from *Santa Margarita* to San Luis Obispo. The convention wound up making an offer, and the delegates left encouraged. A short time later San Luis Obispo celebrated the biggest and best Independence Day in its history. Pat Murphy was president of the occasion and A.R. Estrada was standard-bearer. Although an eloquent speaker, the general limited his remarks to a few words. At dawn a cannon burst forth giving the signal to everyone in town to raise the American flag. During the day there were parades, bands playing and horsemen dressed in appropriate regalia.

It was a long time coming, with ups and downs, delays and disappointments, but on May 5, 1894, the first train arrived in San Luis Obispo. The people would no longer have to rely on stagecoaches or steamers landing at Avila. Pat was among the dignitaries invited to join the Southern Pacific's special train north of the Cuesta. San Luis Obispans had never seen the likes of the celebration that started with a cannon blast to announce the special train bringing four hundred or more guests who thrilled at the panoramic view from the railroad's winding course with its hairpin turns. That evening all gathered in the ballroom of the Ramona Hotel dressed in their finest to take part in the grand march. General Pat was among the guests in one of the "supper rooms" to which special guests were invited.

As it was with most of the large land grant owners there were good times and bad, and for Don Patricio the bad outnumbered the good. He lost the *Margarita* in 1900 to a San Francisco family named Reis. The story told by Pat's relatives was that he was engaged in a poker game at the Pacific Union Club, was drinking heavily and, at the turn of a card, the *Margarita* passed from Murphy hands. According to the Reis heirs, this was not the case. They said it was a foreclosure.

The story of General Patrick Washington Murphy took an ironic twist when a grandson of his brother Bernard uncovered a little black box in his father's law office. A casual examination revealed manuscripts, family records, and a little bankbook of the Hibernia Savings and Loan Society, which, in the intervening years, had become the Hibernia Bank. The account had never been closed. After considerable legal maneuvering, a grandnephew of Patrick Murphy, representing the heirs of Pat's brothers and sisters and Mary Kate O'Brien's heirs, and the Hibernia Bank settled for ninety-five thousand dollars.

CHAPTER TWENTY-FIVE

Lady Diana Murphy Hill Rhodes

The most visible of the Murphy women was Dan and Mary Fisher's daughter, Diana. Not only was she was one of the most beautiful women of her time, but her glamorous life style was widely recorded in the press. Because of her father's several million-acre ranch in Durango, Mexico, newspapers frequently referred to her as the Duchess of Durango.

Lady Diana Murphy Hill Rhodes was one of the great beauties of her day. Unlike her earthy Murphy relatives, she was socially ambitious and achieved her goal when she married Sir George Rhodes, secretary of the British Admiralty. COURTESY CALIFORNIA HISTORY SECTION, CALIFORNIA STATE LIBRARY.

Dan had wanted a boy when Diana was born in 1859 at Ivy Farm, the newly completed family mansion in San Martín. He compromised by calling her Dannie. Dan doted on his pretty little daughter and indulged her to an extreme. Two years later, when their sixth child was born, he finally got a son. They had named the infant son they lost Martin Daniel, so in keeping with a family tradition of naming one son for the patriarch, they called their son Daniel Martin.

Although Diana and her father were close, she was bored with life on her parents' *San Martín rancho* where they spent most of their time. According to her cousin, Wilhelmina Murphy Bassett, she once said to her mother, "Must I always have to hide my pretty face behind these trees?"

This attitude was apparent at Notre Dame Academy where she wasn't the most popular of the many Murphy girls to attend the school. When she left after two years, the sisters shed no tears. They felt she was distracted by what her father's money could buy and by her belief that her beauty was being wasted behind the schools walls. She subsequently enrolled in St. Mary's private girls' school in Oakland and later attended Madame Brosch's in San Francisco. Despite her lack of academic interest she became fluent in Spanish and French and an accomplished pianist.

With her large violet eyes inherited from her Irish father's side and her black hair and olive skin from her Spanish grandmother, Liberata Ceseña Fisher, she was a striking beauty. Although she had many suitors, it was the tall, handsome, fair-haired Hiram Morgan Hill who captured her hand. Orphaned at an early age, he and his sister, Sarah Althea Hill, were raised by relatives in Cape Girardeau, Missouri by relatives who, when they heard rumors of a romance with a first cousin, suggested Morgan move post haste to San Francisco where he had a grandmother. His sister,

Sarah Althea Hill, elected to accompany him. He worked at a bank in San Francisco and, on the side, modeled for Bullock and Jones, a fashionable men's furnishings store.

The couple met in Santa Cruz at the elegant Pope House where the likes of the Crockers, Stanfords and Fairs came to enjoy the fresh sea air. Adding to their glamorous image, Morgan, as he preferred to be called, and Diana presented a handsome picture driving in his Brewster buggy with a matched pair of trotters down the willow-lined Alameda in San Jose.

Mary and Dan Murphy were unhappy over Morgan Hill courting their daughter, and when Diana told her father that Morgan Hill wanted to marry her, Dan was outspoken in his disapproval. Her parents may have had several reasons for objecting to Morgan. First there was the age difference. He was thirty-three and she had recently celebrated her twenty-second birthday. Too, it could have been because of religion. Both the Murphys and the Fishers were devout Catholics, while the Hills were of Quaker persuasion. Added to this Dan, who had accomplished so much at an early age, was impatient with Morgan's being a mere bank clerk and a male model and didn't approve of Morgan's running with a "wild" crowd.

Morgan's courtship of Diana provided grist for the gossip mills. Whispered about was a story that Morgan had proposed several times to Diana, but that she was playing it coy. In truth it was probably indecision because of her concern of going against her parents' wishes. About this time Morgan met an attractive young woman visiting San Jose named Kate Stevens, daughter of a Michigan lumber king. Morgan confided to Kate, who he knew was engaged to another, of his love for Diana and of her procrastination in giving him an answer to his proposals. Together they contrived a little plot. Morgan was to park his Brewster buggy frequently in front of the Stevens's house—anticipating this would stir up rumors that Morgan had switched the object of his affections.

This did the trick. About the middle of July, Diana and Morgan were among a party of young people who left San Francisco for Santa Cruz on the yacht *Nellie*. At the Pope House where they first met, Diana asked Morgan if he still wanted to marry her. Naturally, he said, "Yes." On their return to San Francisco Diana and her bosom friend and confidante, Mattie George, checked into the Palace Hotel. A few days later they met Morgan and a friend at city hall. With them was the Rev. Dr. Jewell, pastor of the Howard Street Meth-odist Episcopal Church. A clerk escorted them to a room where the clergyman, with the understanding that it be kept secret, performed the marriage ceremony. Diana took the next train for San José to join her mother, but not a word to her of the marriage. Mary apparently never suspected even though Morgan visited almost every Sunday.

Hiram Morgan Hill (he never used Hiram) was the first husband of Diana. As her father, Daniel Murphy was dying, he begged her not to marry Morgan Hill. She and Morgan, however, had been secretly married in San Francisco. COURTESY BETH WYMAN.

Diana's trip to Nevada in December of that year to visit her father turned out to be the last time she would see him. He had gone out in a snowstorm to rescue his cattle, took a chill and came down with pneumonia. As his condition worsened Diana and her mother boarded a train in San Jose for Elko. Her brother, Daniel, was in the East in school and couldn't make connections in time to see his father alive. While still lucid, Dan talked to Diana about Morgan Hill and begged her not to marry him. Although ridden with guilt because she had already married Morgan, she gave her father her promise. Nevertheless it continued to prey on her conscience and she contemplated getting an annulment until her friends talked her out of it. Nevertheless, it seemed to affect her persona and that of Morgan.

He insisted that she make known their marriage and when she refused, he seemed to fall apart. Diana, hit hard by the loss of her father, entered a period of dissipation and flirtation with the local lads. She visited Sacramento during a session of the legislature where her colorful, fun-loving cousin, Patrick Murphy, was a member of the assembly representing San Luis Obispo County. Knowing nothing about her secret marriage, he introduced her about. Local gossip had it a young senator from Southern California had the inside track and would lead her to the altar at the close of the session. For obvious reasons it was just gossip.

Finally, unable to salve her conscience, Diana consulted with a prominent San Jose attorney, S.F. Leib. She told him of her false promise to her father and that his dying words haunted her. Bluntly she told Leib she wanted a divorce. He hired a San Francisco detective to trail Hill. In about a month he filed a complaint form charging adultery. Before the case could be heard, Diana's friends again interceded by persuading her to reconcile.

She and Morgan went to Europe for an extended honeymoon. When they returned they built a country estate called Villa Miramonte on her share of the nine thousand-acre land grant *Ojo de Agua de la Coche* she had inherited from her father. Built of lumber hauled from Watsonville by oxen-pulled wagons over primitive mountain roads, it featured mirrored walls, twelve-foot-high ceilings and fireplaces of Minton tile in the parlor and dining room. It was here their only child, Diane Murphy Hill, was born.

Much of the time they hobnobbed with the social elite of San Francisco. The fun days lost their luster, however, when Morgan's sister, Sarah Althea, a golden-haired beauty with luscious black lashes that set off her bright blue eyes, became embroiled in a juicy scandal involving her love affair with Senator William Sharon. He was known as the King of the Comstock Lode, not only for the fortune he had made in silver but for his ruthless business practices, particularly those involving his partner, William Ralston,

founder of the California Bank. At Ralston's mysterious death, the Palace and Grand Hotels and Ralston's enormous estate at Belmont went to Sharon. Nevertheless, he and Sarah Althea got along just fine for fifteen months until she discovered the senator had been unfaithful.

Influenced by the notorious and mysterious Mary Ellen "Mammy" Pleasant, a light-skinned African American woman and housekeeper for banker Thomas Bell, whom she had first known as a child on her family's Girardeau plantation, Sarah Althea signed a complaint. In it she swore Sharon was her husband according to terms of a marriage contract that had been drawn up in his office in compliance with the Civil Code and which he signed before she moved into her quarters at the Grand Hotel. She claimed that he had deserted her and accused him of adultery, naming nine women. In the Sharon v. Sharon suit she asked for alimony and a several million dollar settlement.

As a consequence of it all, Sharon was arrested. So infuriated was he that he counter-filed with a suit in the Federal Circuit Court, called Sharon v. Hill, for the purpose of having the marriage contract declared a fraud.

Altogether the press had a field day. Dubbing Sarah Althea the "Rose of Sharon," they noted among other things that no relatives of Sarah Althea were in the courtroom to lend their support. Morgan was apparently humiliated at this stain on the family name. He and Diana were now shunned socially, and no longer among the "beautiful people."

The trial was under way for several weeks before the notorious "Mammy" Pleasant appeared in the courtroom. She continued to be there every day after that. Sitting behind the plaintiff and her counsel, David Terry, she conferred with them in a whisper. Although she first denied it, Mammy had a financial interest in the case. She had put up fifty thousand dollars of her own money, expecting to reap a large award. The hypnotic gaze of her one blue and one black eye and her reputation for practicing voodoo, could put fear in the heart of such as Senator Sharon. She could appear to be generous and kind, but she could also be unscrupulous and manipulative.

The gorgeous, golden-haired Sarah Althea titillated the waiting crowd outside the courthouse as she alighted from her carriage each day wearing a different stunning costume. Mammy's attire was always the same—a black dress with accents of white at the throat and cuffs, and a scarf over her black hat that tied under her chin. The stories of Mammy Pleasant's intrigues are replete with suits and counter-suits, plots and sub-plots, of illegitimate births, of slander and perjury. She even took credit for having willed the death of Senator Sharon. He did pass away in one of his rooms at the Palace Hotel nine months before completion of the Sharon v. Hill case.

The legal process reversed itself. First the Superior Court had ruled supporting the legality of Sarah's marriage to Sharon and the Supreme Court affirmed it. But when Sharon's heirs appealed the case was referred to the United States Circuit Court for a hearing. The opinion handed down declared the marriage contract to be a forgery. In turn, the appeal by Sarah Althea's attorneys was rejected by California Supreme Court Justice Stephen Field. The justice held a strong dislike for Judge Terry, whom he succeeded on the bench. In a duel, Terry had killed Field's friend, U.S. Senator David Broderick. The case was now closed, but it was not the end to the traumas of Sarah Althea. For unseemly conduct in the courtroom she and Justice David Terry, whom she had since married, both received jail sentences.

But fate again intervened. In a train station between Fresno and San Francisco, Judge Terry and Sarah Althea had a chance meeting with Justice Field. In the confrontation that followed, Terry slapped Field in the face and was quickly shot and killed by a marshal guarding Justice Field. The marshall, David Naegle, recently from Tombstone, Arizona, was acquitted. This was the last straw for the fragile Sarah Althea. Showing brotherly concern, Morgan Hill came to Stockton to support her through the funeral service.

Afterward, she joined Mammy Pleasant in San Francisco. Mammy apparently had plans to capitalize on the publicity of the trial and on Sarah Althea's exceptional beauty by launching Sarah on a stage career. But Sarah's mental health deteriorated rapidly. As a consequence, Mammy committed her to the state asylum for the insane at Stockton where she remained until her death forty-five years later. During that time Mammy Pleasant never visited her protege nor even so much as wrote a line to her. At the age of 89, Mammy died—friendless and, it was said, practically penniless.

To get away from the incessant trial publicity that plagued them, the Hills took off for Europe where they sojourned for several months. When they returned, Morgan and Diana, who had been drifting apart, agreed to separate. Morgan moved to Nevada to manage his late father-in-law's vast estate and for some

time did an excellent job of holding it together. As time went on, he became increasingly eccentric. Once a dandy and a natty dresser, he went about in a dirty work shirt, a slouch hat and bib overalls—one pant leg tucked into a boot and the other hanging outside the boot. He would travel from Halleck to Elko by horseback or carriage. On occasion, if he took the train, he carried a dilapidated valise. A source of amusement to the locals in Elko, when he arrived he would have a Pullman towel draped around his neck, another protruding from a pocket. The porter helping him off the train would pretend not to see as Morgan handed him a silver dollar tip.

Back in California the *Ojo de Agua de la Coche* property Diana had inherited from her father was sold in 1892 to C.H. (Chauncy) Phillips, the largest and most respected land developer in the state. About that same time Phillips was working with Diana's cousin, Patrick Murphy, in promoting the Southern Pacific Railroad to run its tracks from Patrick's *Rancho Santa Margarita* to San Luis Obispo. The two men were also working on the development of San Luis Obispo.

The following year, lots of five, ten, twenty acres and upward were offered for sale at Morgan Hill, the name Chauncy Phillips had given to his new town. The name was probably a carry-over from the period when the Hills' friends would come down from San Francisco for a visit and tell the conductor they wanted to get off at Morgan Hill's. Other people thought the name came from the prominent peak that loomed in the background. Actually it was still being called Murphy's Peak and would continue to be for another twenty or so years, at which time it reverted back to its original Spanish name, *El Toro*.

In the meantime, Diana's brother, Daniel M. Murphy, following her example, decided to sell the adjoining San Martín property he had inherited. It had become too valuable for wheat farming so he, also, sold to Phillips. According to a *San Jose Mercury* account, the eight thousand-acre parcel represented only one-sixtieth of the land he inherited from his father. By this time, he was frittering away most of his fortune on whatever was the whim of the moment—including trips around the world. On one occasion he cabled his overseer, "sell one thousand head of cattle and send me the money at once."

Then there was his hunting lodge at Coyote Dam where he raised thoroughbred horses and registered dogs, providing kennels for the dogs luxuriously furnished with specially constructed beds. On the stable doors of his thoroughbred horses were their portraits, painted by an English artist. He sold his cattle and land in White Pine County to a former partner of his father's, A.C. Cleveland.

Two years later he sent his uncle, Fiacro Fisher, with Tom Columbet, Juan Feliz and his *vaqueros* to herd one hundred head of saddle horses to Duck Valley to be sold at the Duck Valley Rodeo. Fiacro also worked on Dan Murphy's Durango ranch. On one occasion he brought his son, Robert, down from the Coyote Ranch and while working the iron mine, the young man was killed.

Shortly after the sale of the San Martín land, young Daniel left with his agent for Durango to check out a proposed sale of part of his four million acre inheritance. Durango seemed to bring bad luck for the Fisher-Murphys. While attending a fiesta there a celebrant, who had imbibed too much tequila, hit young Daniel on the head with a bottle. He became mentally incompetent until his death came in 1915.

After her separation from Morgan Hill, Diana maintained a residence in San Francisco and another in Washington, D.C. where she leased the British Legation building. With her taste and flair she turned it into a showplace where she entertained in grand fashion the likes of President Teddy Roosevelt and the diplomatic set. She soon became the Pearl Mesta of her day. From time to time Diana would take her adored little Diane to Elko to visit her father on the ranch. After attending the finest private schools, Diane's debut in Washington society, widely recorded by the press, was an elaborate affair befitting the daughter of the capital's leading socialite. A short while later the pretty debutante sailed for Europe where she met the slim, handsome Baron Hardouin de Reinach-Werth of France. Although she had never shown interest in social status, Diane and the Baron became engaged— much to the delight of her mother who looked forward to having a title in the family.

Their wedding in 1911 took place in St. Matthew's Cathedral in Washington, and according to the *Washington Post*, was "witnessed only by the parents of the bride." Afterward, the couple sailed to Europe for an extended honeymoon and a visit with his parents in France. They were planning to leave for the ranch he inherited in Manitoba, Canada when she received a communiqué in London from her mother with the news that her father had suffered a stroke. It had become apparent that Morgan, who was now deaf, was deteriorating mentally as well. The newly married cou-

ple changed their plans so they could first visit her father in Elko. This was not to be. Diane became ill and was hospitalized with a nervous breakdown in a sanitarium near Regent's Park. When the nurse went into another room to get something, Diane jumped from the balcony falling on to the cobblestones eighteen feet below and suffered a fatal skull fracture. The strain of mental instability that ran through the Hill family, affecting first her aunt, Sarah Althea, then her father, Morgan Hill, may have doomed Diane Murphy Hill de Reinach-Werth.

By a strange twist of fate Morgan, who died in 1913, was buried in the Santa Clara Mission Cemetery in a grave next to his father-in-law, Dan Murphy, the man who had so opposed his marriage to Diana.

Crushed at the loss of the daughter she loved so much, Diana left immediately for London to assist her son-in-law in making burial arrangements. With Morgan gone, she asked Captain Werth, as he preferred to be called, to go to Elko with her to help handle the Nevada properties. The dashing, slender young Frenchman entered into the ranching life of Elko, making many friends. He could be seen in his sports car driving to the *Rancho Grande*, followed by his two hounds; they would run part of the way and then jump into the back seat of the car to ride the rest of the way.

Diana built a home with an apartment for the captain and one for herself with a common living room— a life style that caused Elko residents to raise their eyebrows. In this elegant setting they entertained with gourmet dinners. She usually served duck rare with the blood running out. An Elko woman, a widow with three children, who helped in the kitchen, said most of it was left on the plates—that there was so much waste of food. Yet when she asked Diana if she could take it home to feed her children, Diana refused, saying that such action might lead to thieving. She said she didn't want to set a precedent.

When the war broke out in 1914, Captain Werth was called into service by the French government. Before sailing for France he bought a boatload of horses to be used by the French in the war and had them shipped from New York. After the war he married a French actress and that was the last Diana or any of his Elko friends heard of him.

Diana missed the captain and the brightness he brought into her life. Without his companionship, Nevada cattle country living was not her cup of tea. At the end of World War I, Diana was sixty-two years old, still beautiful and retained her graceful bearing,

She moved to London where she met a British baronet, Sir George Rhodes, undersecretary of the British Admiralty. He was a widower with three children and was first cousin of Sir Cecil Rhodes, the "Empire Builder" and founder of Rhodesia. They were married in St. Mary's Catholic Church, London. Following the ceremony, the groom's son-in-law, Ian MacPherson, who held an important post in Lloyd George's cabinet, gave a wedding breakfast for them attended by a small but impressive list of titleholders. Thus it was that little Diana Helen Murphy Hill of English, Mexican and Irish background, who started life on a far away California cattle ranch, became Lady Diana Rhodes and was presented to the queen.

After only a year and one half of happy married life, death came to Sir George. A saddened Diana moved into a palatial villa at Cannes on the French Riviera. There she mingled with the smart set, as reported in the press, including the Duke of Windsor (Edward VIII). She amused herself nightly playing roulette at Monte Carlo—was said to pay off her losses with parcels of her Nevada rancho lands.

In 1937, the curtain descended on the glamorous life of Lady Diana Murphy Hill Rhodes. At her bedside in her villa at Cannes were two of her late husband's children, the Baroness MacPherson and Lieutenant Colonel John Rhodes. She was buried in the Anglo plot of the Catholic cemetery at Cannes. Anglo-American society leaders of the Riviera present to pay tribute to this remarkable woman included her son-in-law, Viscount Ian McKenzie, who was Great Britain's undersecretary for war, and Prince Gennaro de Bourbon, cousin of King Alfonso of Spain with his princess.

She was far away from the Murphy family plot in the Santa Clara Mission Cemetery where her parents, her brother, and her first husband, Morgan Hill, are buried.

CHAPTER TWENTY-SIX

The Golden Wedding Anniversary

On a late spring afternoon in 1881 Martin and Mary Bolger Murphy's five children and their children, now numbering twenty-two, came to Bay View for Sunday dinner. Although happy as always to be coming home, this was more than a casual visit. Their parents would be celebrating their fiftieth wedding anniversary in a few months. Eager to talk about plans were their daughters, Polly from San Francisco with her husband, Richard Carroll, and Nellie with her husband, Joaquín Arques from their Santa Clara ranch. Joining in the discussion were the Murphy sons: Barney or B.D. as the family called him with his wife Annie McGeoghenan from San Jose; Patrick from *Rancho Santa Margarita* and Jimmy, the youngest and the only child born in this house, with his wife, Wilhelmina, from their Milpitas ranch.

The younger children, who always enjoyed coming to their grandparents' house, were let out to play tag on the broad expanse of lawn in front of the house, climb the gnarled old pepper tree or, the most fun of all, play hide-and-seek under the traveling Smyrna fig tree, whose branches, extending over a ninety foot area, had reached down and rooted themselves to the ground much like a Banyan tree.

Meanwhile Polly, intent on getting the discussion under way, began, "Of course, we should have a big party, but Nellie and I are wondering what to do about the guest list."

Before she could continue Martin, with a knowing smile, raised his hand to interrupt her. "Excuse me, my dears," he said, "but you are not to worry. Your mother and I have talked about this. We were afraid of leaving someone out, so we have decided to invite everybody."

His daughters' eyes widened in amazement. "How could you do that?" they asked in unison.

"Very simple" he said. "We'll place an open invitation in the newspapers."

Mary, who had been quietly listening until this point, by way of explanation, said, "We're planning to charter special trains to bring guests from San Francisco and San Jose."

Martin and Mary Murphy's Golden Wedding Anniversary at Bay View was attended by between five and ten thousand guests and said to be the biggest bash California had ever seen. A letter of congratulations came from Sister Marie Catherine of Notre Dame. COURTESY MARJORIE PIERCE.

Nellie's husband, a genteel Spaniard who had lived in Chile, had been listening intently. He shook his head in wonderment. Thinking aloud of the magnitude of such an event, he said, "I suspect you'll be having a lot of people. I mean a really whole lot of people." Hesitating for only a brief moment, he continued, "Would you let me handle the food and beverage arrangements?" Mary and Martin looked at each other in surprise, then nodded their heads in agreement.

A lot of people did indeed come. California had never seen the likes of it. On that morning of July 18, 1881, the press estimated that between five and ten thousand people (most settled on seven thousand) descended on Bay View Ranch. By eight in the morning jubilant partygoers in carriages, buggies, farm drays and even some on horseback crowded dusty El Camino on their way to Bay View Ranch. By noon some eight hundred vehicles were counted on the grounds.

Starting off the festivities, Parkman's eighteen-piece band serenaded the guests of honor as they walked out the door of their handsome twenty-room house which Martin had shipped around the Horn from Maine and assembled on the property some thirty-two years earlier. The crowd that gathered around gave the couple a rousing cheer before escorting them to the one hundred foot-long, architect-designed dance pavilion where they took their places seated at one end under a canopy of flags wreathed with flowers and evergreens. Mary, wearing a simple gray silk dress and a Victorian style cap with a white lace ruffle that framed her face could scarcely conceal her excitement while Martin, still sturdy and erect for his seventy-four years, his brown hair and sideburns sparsely sprinkled with silver, accepted it all with his usual composure. At one side of the pavilion was a stand for the musicians and surrounding it were chairs for guests.

When the first four-car excursion train from San Francisco arrived at Murphy Station, the band was on hand to greet the celebrants as they disembarked and were led to the pavilion where they stood in line to pay their respects to the honorees. The couple's heads brimmed over with memories as they greeted friends from their native County Wexford, from their farming days in Canada, and still others who had been with them at Irish Grove on the Missouri River. They were especially pleased to see their companions from the long and memorable crossing of the plains thirty-seven years earlier. They included Dennis and Patrick Martin, John and Michael Sullivan, Moses Schallenberger, and from Mayfield Farm, the former Sarah Montgom-

ery with her current husband, state Senator Joseph Wallis. Sarah had become active in the woman suffrage movement after her first husband, Allen Montgomery, deserted her in San Francisco and her second husband, the handsome, charming Talbot Green, who at the time was running for mayor of San Francisco, turned out to be an imposter named Paul Geddes and was apprehended by the law.

The trains from San Jose kept coming until the streets of the town were left virtually empty. So great was the excitement that at noon the San Jose Board of Supervisors adjourned, banks closed and, although in the middle of a trial, the esteemed Superior Court Judge David Belden, a close friend of Martin, dismissed the court for the rest of the day. He was just as eager as the sheriff, jury, witnesses and counsel to attend the party.

Never before in California had so many prominent persons attended the same social function. The Honorable William Gwin, California's first full-term senator, arrived on the train, as did many San Francisco financiers, friends and business associates of the hosts. They included some of the wealthiest men in California, many of whom were Irish, who, like the Martin Murphy, Jr., had a hunger for land. Like Martin, they satisfied that hunger with a feast. James D. Phelan, who parlayed his fortune acquired from selling merchandise to the miners into a real estate empire, attended with his twenty-year old son, James. The younger Phelan was destined to become mayor of San Francisco and a U.S. senator noted for his philanthropy and as a patron of the arts.

Another Irishman who came with the Murphy-Stephens party and made it big in real estate was John Sullivan. After a successful trip to the Mother Lode, he started buying up San Francisco lots. At the request of Bishop Alemany, who was concerned because so many of the poor lost money in bank foreclosures, he organized the Hibernia Bank and served as its first president. Always called "Big John" because of his size, he shared grandchildren with Jim Phelan. His son, Frank, who became a state senator, congressman, park commissioner and attorney, married Phelan's daughter, Alice.

Still another of the state's wealthiest men (with a fortune estimated at five million dollars) was San Franciscan Peter Donahue who had a summer home called Laurelwood Farm on the Ulistac land grant near the Murphys. An industrialist with a long list of "firsts," he introduced gas lighting to San Francisco, built the first gas works and the famous Union Iron Works. His wife, Anna Downey, sister of California's Civil War governor, John Downey, accompanied him to the party but not in her glass carriage which always created such interest when she traveled in it around San Francisco.

Guests moved about the park-like grounds, admiring Mary Bolger's well-kept gardens with their beds of fragrant Castilian roses, stock, passion vine and other flowers. Turning toward the east they saw vineyards and orchards, horses and cattle grazing and beyond, almost as far as one could see, fields of golden wheat, swaying in the gentle breeze. One newspaper writer named Charles South described the house as "standing out like a white ship in a sea of gold."

In the mixed crowd, bankers, businessmen and politicians, attired in their smartly tailored suits, contrasted with the more casual dress of the farmers and farmhands (some in bib overalls) whose faces were lined from years in the valley sun. Alfred Doten, from his nearby farm, had many anecdotes to tell about the Murphys, including those about Martin's stable of fine horses and his fun in having horse races on the ranch. He tells also about the time the four older Murphy boys came to get him to play his fiddle for a dance at the house.

Many of their early California friends came such as the Arguellos from the *Las Pulgas rancho*; Doña Juana Briones, one of California's best-known women of her time for her nursing of the sick from neighboring *Rancho La Purisima Concepcion*, and Segundo Robles, a regular Sunday dinner guest of the Murphys, who rode his horse over from his *Rincon de San Francisquito Rancho*. Robles, a former major-domo of Mission Santa Clara, played an important part in the discovery of quicksilver.

For the first time in thirty-five years Don Segundo met attorney, banker and former speaker of the assembly, C. T. Ryland. They laughed as Ryland recounted the day in 1849 when, after an unsuccessful stint in the gold fields, he was walking from San Francisco to San Jose. When he came to the San Francisquito there were so many wild cattle he decided he needed a horse. Robles provided him with one and told Ryland when he reached the pueblo to turn the horse in the direction of home and then give him a slap on the rump.

Providing splashes of black and white in the colorfully dressed crowd were the Jesuit fathers from Santa Clara College, many of whom frequently celebrated Mass at Bay View for the devout Murphy family. One of the Murphys' special Jesuit friends was young Father

Robert Kenna, brother-in-law of John Sullivan. He became one of Santa Clara College's outstanding presidents. As a result of the eloquent talk he gave before the state assembly he played a key roll in Andrew Hill's fight to save the Santa Cruz Mountains old-growth redwoods. Another was Father Nicholas Congiato, pastor of St. Joseph's, who built the beautiful church in San Jose. This church would, within a little over one hundred years, become a cathedral and then a basilica.

Caius Tacitus Ryland also attended. FROM *CLYDE ARBUCKLE'S HISTORY OF SAN JOSE* BY CLYDE ARBUCKLE, 1985.

Representing the Breen family, members of the Donner Party who had stayed with the Murphys at their Cosumnes ranch when they came down from the mountains, was Judge James Breen. A college friend of Barney Murphy's at Santa Clara, he succeeded him as president of the school's alumni council.

The black-garbed nuns from College of Notre Dame, who revered the Murphy name for the important role the family played in the founding of their school, were warmly greeted by former students that included most of the Murphy women. The big surprise and joy for them, however, was to be greeted by John Townsend, son of Dr. John and Elizabeth Townsend, leading members of the Murphy-Stephens wagon train who were victims of the cholera epidemic of 1850. At the age of three young John was Notre Dame Academy's first boarder and only boy. He did yeoman duty assisting Uncle Ike Branham with the barbecue.

Joaquín Arques skillfully handled the logistics. True to Murphy tradition he planned generously. In addition to the beeves from Martin Murphy's prized herd, there were dozens of sheep and swine, wagonloads of chickens, hams and headcheese; a thousand loaves of bread, and salad served from bushel baskets. To wash all this down there were five-hundred gallons of coffee, fifteen half-barrels of lager, hogsheads of lemonade, casks of whisky, a carload of champagne and the finest wines.

Isaac Branham celebrated with the Murphys and oversaw the barbeque. FROM *HISTORY OF SANTA CLARA COUNTY, CALIFORNIA, ALLEY, BOWEN & CO.,* 1881.

Captain Isaac "Uncle Ike" Branham, an old friend of the Murphys from the *pueblo* days, came up from Southern California and, assisted by a corps of butchers, carvers and cooks, lived up to his reputation as the premier master of the barbecue in the state. Described later by the press as "well larded in person and rosy cheeked," his movements were deliberate as for six hours he moved up and down the one hundred fifteen foot long, four foot wide, and four foot deep pit with a small mop basting the meat from a can containing his secret recipe.

At noon he took off his apron and the carvers took over. He was free to visit—first with Martin's brothers, John and Dan. John, like Uncle Ike, had been active in the pueblo government after the United States takeover. Dan, who also held political offices, had come over from his Elko, Nevada ranch with his daughter, Diana. He reminded Uncle Ike of their serving together on the San Jose and Santa Clara Horse Railroad Association board. They laughed together about the fox hunts, California-style, Ike used to stage from his ranch south of the *pueblo*—especially over one that got away on a Christmas morning.

They remembered that no fox was available that day so the scent of a coyote that had been inadvertently killed by the hounds was substituted. The lead rider sped off, followed by hundreds of baying hounds, the huntsmen and carriages filled with frightened women dressed in their finest, hanging on to their seats with one hand and their hats with the other. For some reason, instead of taking to the countryside, he headed north on Monterey Road and onto Market Street, arriving in front of St. Joseph's Church just as eleven o'clock Mass was letting out. The startled parishioners gazed wide-eyed at the sight and sound of the yipping hounds, of the pounding of the horses' hooves and of the cries of the distraught women in the carriages.

Emotions ran high for Moses Schallenberger when he met Dennis Martin, a fellow member of the Murphy-Stephens party. Moses recalled for those standing nearby how Dennis had saved his life when he courageously hiked alone and through deep snow without a path to follow or a landmark, to rescue him at Donner Lake that freezing December of 1844.

Then Moses and John Murphy laughed with John Sullivan, who had joined the group, over the trick they played on him one night on the trail when they were passing through Indian country, and they let his cattle out, yelling, "Indians!" Sullivan remembered the incident well, saying that it didn't seem funny then—that he had worried all that day that they might be attacked by Indians. Moses, a pal of the Murphy boys during the crossing whose friendship continued afterwards, kidded Barney Murphy, San Jose's mayor, who was only three at that time, about his carrying him half way across the country on his shoulders.

With great fanfare, at twelve-thirty in the afternoon the band struck up "All Haste to the Wedding," the signal for the bride and groom to be escorted to the table by their children. The decorations, a surprise to Mary, featured at one end of the table a large horseshoe, fashioned of fragrant orange blossoms symbolic of California. A stuffed bear represented the forty-niners and the Gold Rush which started them off on their road to financial success. Hanging from an oak branch overhead was an enormous wedding bell of white flowers with the figures -1831- inlaid in red rosebuds on one side and on the other -1881. Standing beside the table was a seven foot tall replica of the dance pavilion created of sugar and nuts, topped with the figures of a bride and groom, and placed in front of them was an elaborately decorated, many-tiered wedding cake.

It was a busy day for Uncle Ike Branham. He did double duty by acting as master of ceremonies as well as chef. He first introduced Senator William Gwin, a friend of his since the meeting of California's first legislature at Uncle Ike's *adobe* on the *plaza* in the *pueblo* in 1849. Gwin was elected California's first full term senator at that meeting. He extolled Martin's integrity, saying he was never known to have foreclosed on a piece of property and always charged below the going interest rates.

Next on the agenda was Peter Donahue, a partner of Martin's in the purchase of the *Santa Margarita rancho* and one of three builders of the San Francisco-San Jose Railroad. Donahue related how Martin gave the right of way for the railroad to come through his property, but, he added, with the requirement that there would be two flag stops at the ranch for the convenience of family and friends and that he be given permission to build Murphy Station. What's more, he added, Martin received a fifteen-year pass.

Adding his few choice bits of wit was Martin's old friend, the elegant and charming Tiburcio Parrott of San Francisco's premier banking family and one of the first gentleman farmers to grow grapes and produce wine at St. Helena. He was followed by J.J. Owen, publisher of the *San Jose Mercury*, representing the press. He read an essay from his collection of "Sunday Talks" and presented the compilation to the Murphys.

It was now the family's turn. General Patrick Murphy, the couple's eldest living son, with his gift for words, eloquently eulogized his parents, saying they lived a life of love and charity and that his father, like his father before him, always led morning and evening prayers. "To some" he said, "father might seem reserved, but when we were little he would get down on the floor and play with us. At Christmas when we children came down the stairs, he would be there to greet us. Putting our hands together, he would slip in a gold piece and say *Feliz Navidad*.

Describing his mother he said, "She was thrifty, but never parsimonious, generous and honest to the extreme. If a horse or cow was being sold she insisted the buyer know every little thing about the animal, more than they wanted to know." He told of the time a new neighbor, actually a squatter, came to borrow some milk. The hired girl, Hannah, asked if she should take the cream off the top. He said his mother replied indignantly, "You'll give it to him just as the cow gave it to you."

James, the youngest son, read a poem he had written for his parents, and next, a young Aloysius O'Brien stood on a bench to read a poem written by a niece of the Murphys, Sister Anna Raphael Fitzgerald. Other grandchildren read letters of congratulations that included those from Governor George Perkins, Supreme Court justice and California's first governor Peter Burnett, from Burnett's son-in-law, William Wallace, who was to follow in his footsteps on the court and from the Murphys' close friend, Archbishop Joseph Sadoc Alemany, who regretted that a confirmation ceremony would preclude his being present. A room in the house called the Bishop's Room was set aside for him when he traveled to churches in his far flung diocese.

When called upon to speak, the Honorable C. T. Ryland, former judge, speaker of the house and attorney, rose to the occasion. "Ladies and Gentlemen" he said, "I drink to the health of the bride, Mrs. Martin Murphy, whose good sense and knowledge of justice beat all the lawyers and judges of this land in getting the squatters off this ranch...I desire to show that in her dear old head and heart she has the sense and sentiment that outstrips the law in the race for justice. He told about a time when Martin and a corps of lawyers including Judge William Wallace were trying all the machinery of the courts and making use of all appliances known to the practice of law to clear the ranch of squatters.

"Mrs. Murphy," he said, "with a higher sense of justice and nobler sentiments of humanity was feeding and ministering to the necessities of the families of the very men who were fighting her husband in the courts." The sequel, he said, "was that the squatters were ashamed of themselves, lay down their opposition at her feet and abandoned the ranch. She secured with kindness what the hand of law was trying to accomplish."

The guests spontaneously rose to their feet and bowed their heads in respect. Stepping forward to add his contribution, Patrick Fenton, a friend and neighbor from his nearby Fenton's Indian Mound Ranch on the Ulistac land grant, told of the time their son, a seminarian student at Santa Clara College, came home for a holiday and went down to see some Indians living on the ranch. Apparently, drunk, the Indians killed him. "Mary Murphy" he said, "was the first to arrive to give us moral support. She knew what it meant to lose a child. She had suffered the loss of six children herself." He then added, "Another time, when Indians set fire to our barn, wagonloads of wheat, a year's supply, arrived from the Murphys."

Anna Maria Bascom (always known as Grandma Bascom), at whose boarding house in the *pueblo* Martin Murphy stayed on his early trips to the valley to buy cattle, joined in the chorus of praise for Mary. She repeated a story of when the Murphys were living on the Cosumnes—that Mrs. Murphy's kindness to the Indians was repaid when there was an uprising and even though her husband, the chief, lay dead nearby, his wife positioned herself in front of the doorway and slept there all night.

At this point Martin, concerned about all those still to be served, rose to thank his friends for their generous toasts and one and all for coming. He regretted that his brother Jim could not have lived long enough to be present for this occasion—the absence of his sisters, Mary Miller, who was ill in San Rafael and Helen, who was in mourning in Stockton for her husband, Charles Weber. But he asked his brothers, John and Dan, and his sisters, Margaret Kell and Johanna Fitzgerald, to join him in offering a toast to someone dear to all their hearts. "He was a man of vision and courage, who led us from Ireland to Canada, by water to Missouri, across the pathless plains and, finally, over the Sierra Nevada to this fair valley which he liked to call the Promised Land. He gave us a legacy of love and caring and, in turn, was loved by all. Please join us in raising our glasses in tribute to the Irish patriot, our late father, Martin Murphy."

The guests then left the tables so that another thousand persons could partake of the feast. Some moved over to the pavilion for dancing and to be entertained by an Irish piper playing an old-fashioned bagpipe, some to board the trains for San Francisco and San Jose. But most, reluctant to have the party of the century end, stayed on until the wee hours. As night fell, the trees were lighted by torches, and the Chinese lanterns strung throughout the gardens and around the pavilion created a magical effect.

In 1981 the City of Sunnyvale held a centennial celebration of the 50th anniversary party of Martin and Mary Murphy. COURTESY STOCKLMEIR LIBRARY AND ARCHIVES, CALIFORNIA HISTORY CENTER.

In his 1892 *Chronicles* Hugh Bancroft wrote:

Far into the night multitudes lingered among the illuminated groves; for the scene was one of unsurpassing loveliness...one that has never before or since been witnessed in California, and one that will long be remembered.

Among other accolades in the press, the *San Francisco Examiner* described the affair as:

A feast that has never been equalled in point of stupendous liberality and profusion on the coast.

The *New York Weekly Graphic*, in its account of the party, wrote:

It was in all probability the largest entertainment ever given by one person on this continent and is chronicled in the annals of the county as one of the prominent points in the history of California.

Epilogue

Martin never fully recovered from his fall. Three years later, when death from pneumonia came to the pioneer and empire builder, the news of his passing sent waves of sadness throughout the state. Flags flew at half-mast, the City Hall and business houses closed for the day. Mourners filled St. Joseph Church for the funeral mass offered by Bishop Alemany. Father John Prendergast, Vicar General of the Diocese, eloquently recounted the life of Martin Murphy — of his contributions to agriculture, stock raising, education and to the life of the Catholic Church.

The funeral cortege of mourners numbered two hundred carriages; others followed on horseback or walking. When they passed the Convent of Notre Dame, an honor guard of sisters and students stood at attention to pay tribute to their patron whose financial and moral support had made their school possible. As the hearse arrived at the Santa Clara Mission Cemetery, the last mourners were leaving St. Joseph's Church.

The press gave praise to the man and his life.

The *Sacramento Bee* led its story with:

> *Santa Clara's most respected, most popular and oldest citizen…*

The *San Francisco Morning Call* wrote:

> *Martin Murphy widely known for his wealth and generous hospitality is worth of study in these days of shifting occupations and gambling speculations…*

The *San Jose Mercury* expressed its sentiments with these words:

> *Seldom has an event transpired within the city's history that has caused so general a feeling of sadness and regret as the death of Martin Murphy. This was testified to by the large number of people who visited the home of the bereaved family. Many were comparative strangers who, actuated by a knowledge of the good deeds of Martin Murphy, desired to look for the last time on a face the had learned to love.*

Mary Bolger Murphy survived her husband by eight years. The headlines of her obituary in the *San Francisco Examiner* read:

> "Mourned by thousands"
>
> "Death of Mrs. Martin Murphy of Mountain View"
>
> "A Life of Love and Charity"
>
> "One of the Most Noted Pioneer Women of the State"

The story details the life of Martin and Mary Bolger: *The news of her death will be a sad blow not only to the community, but to thousands of friends throughout the State. Her generosity made her name a household word among the poor of the land. Her whole-hearted hospitality caused her to be loved by all who knew her.*

The Martin Murphy estate (estimated at five million dollars) by a pre-arrangement was deeded equally among their three sons, two daughters, and the four children of the late Elizabeth Yuba Murphy Taaffe.

In 1898 their son, Gen. Patrick Murphy, owner of one-sixth of the estate sold two hundred acres of his share of the Murphy Ranch to real estate developer, W. E. Crossman, for thirty eight thousand dollars. He retained only seven acres of the grounds around the family home. Crossman, envisioning an industrial center, referred to it as Sunnyvale, the City of Destiny. How prophetic. In less than one hundred years we know refer to it as the heart of Silicon Valley.

Although vigorously protested by civic and historical organizations, Bay View met its demise at the hands of bulldozers in 1961. A large boulder bears a bronze plaque dedicated to Martin Murphy, Jr. and eight years later the Martin Murphy, Jr. Historical Park was established.

Most of the principals of the Murphy-Stephens party remained in the Bay Area. Dr. John Townsend, first American to practice medicine in California, became a prominent citizen in San Francisco where he served as alcalde. During the influenza epidemic of 1850, while nursing the afflicted, he became a victim of the disease, and unfortunately passed the germ on to his wife, Elizabeth. They died within twenty-four hours of each other. Their infant son, named john, was left in the care of his uncle, Moses Schallenberger, who took him to Mission Santa Clara to be raised by the Jesuits.

In 1848 Elisha Stephens was known as the first settler in Cupertino. Stevens Creek Road (a misspelling) was named for him. He took up a homestead later called Blackberry Farm. In 1862 he decided the area was getting too crowded, and he moved to Kern County where he lived to the age of 84. It was a long time coming, but in 1994, due to the efforts of James J. Rose, Mount Stephens in the Sierra near Donner Lake was dedicated to the captain's memory.

Sarah and Allen Montgomery were members of the Murphy-Stephens party. Allen went to the Hawaiian Islands in 1847 and was never heard from again. Sarah next married Talbot Green who turned out to be an imposter named Paul Geddes. One of the first woman suffragist, she lobbied for women's rights, became president of the California Women's Suffrage Association and married Judge Joseph Wallis who built a fine home for her at Mayfield (Palo Alto) where she entertained Susan B. Anthony.

James Miller, husband of Mary Murphy, who stayed with the women and children at the camp on the Yuba River during the frozen winter of 1844, was a pioneer in education and was considered to be the founder of Marin County. He survived his wife by nine years.

Bibliography

Books and Manuscripts

Arbuckle, Clyde. Clyde *Arbuckle's History of San Jose*. San Jose, Calif.: Smith & McKay Printing, 1986.

Arbuckle, Clyde. *Santa Clara County Ranchos*. Cartography and illustrations by Ralph Rambo. San Jose, Calif.: Harlan-Young Press, 1968.

Arce, Francisco. "Memorias Historicas de Don Francisco Arce." Salinas, Calif.: 1877. Manuscript, Bancroft Library, University of California, Berkeley.

Ault, Phil. *Whistles round the Bend: Travel on America's Waterways*. New York: Dodd, Mead and Co., 1982.

Beal, Richard A. *Highway 17, the Road to Santa Cruz*. 2nd ed. Aptos, Calif.: The Pacific Group, 1991.

Begley, Monie. *Rambles in Ireland: A County-by-County Guide for Discriminating Travelers*. New York: Methuen, 1979.

Beilharz, Edwin A. and Donald O. DeMers, Jr. *San Jose, California's First City*. Tulsa, Okla: Continental Heritage Press, 1980.

Bell, Major Horace. *On the Old West Coast*. Edited by Lanier Bartlett. New York: W. Morrow & Co., 1930.

The Bidwell-Bartleson Party: *1841 California Emigrant Adventure*. Edited by Doyce B. Nunis, Jr. Santa Cruz, Calif.: Western Tanager Press, 1991.

Bidwell, John. *Echoes of the Past about California*. Chicago: R.R. Donnelley & Sons, 1928.

Bidwell, John. *In California before the Gold Rush*. Los Angeles: Printed by the Ward Ritchie Press, 1948.

Bidwell, John, John A. Sutter et al. "New Helvetia Diary." Sutter's Fort, New Helvetia, Alta California. Manuscript diary, 1845-1848, Bancroft Library, University of California, Berkeley.

Bray, Edmund. *"Letters to H.H. Bancroft. Sacramento, Calif.: 1872."* Letters, Bancroft Library, University of California, Berkeley.

Brown, James L. *Dissension in Arcady : The Bear Flag Revolt*. Campbell, Calif.: Academy Press, 1978.

Bryant, Edwin. *What I Saw in California*. Minneapolis: Ross and Haines, Inc., 1967.

Buffum, E. Gould. *Six Months in the Gold Mines*. Los Angeles: Ward Ritchie Press, 1959.

Burchell, R.A. *The San Francisco Irish, 1848-1880*. Berkeley: University of California Press, 1980.

Burnett, Peter H. *Recollections and Opinions of an Old Pioneer*. New York: D. Appleton & Co., 1880.

Butler, Phyllis Filiberti. *The Valley of Santa Clara*. San Jose, Calif.: Junior League of San Jose, 1975.

Carlson, Helen S. *Nevada Place Names*. Reno: University of Nevada Press, 1974.

Carroll, Mary Bowden. *Ten Years in Paradise*. San Jose, Calif.: Press of Popp & Hogan, 1903.

Carson, James H. *Early Recollections of the Mines and a Description of the Great Tulare Valley*. Tarrytown, NY: Reprinted by W. Abbatt, 1931.

Caughey, John Walton. *Gold is the Cornerstone*. Berkeley: University of California Press, 1975.

Cendrars, Blaise. *Sutter's Gold*. New York: Harper and Brothers, 1926.

Chittenden, Hiram Martin. *History of Early Steamboat Navigation on the Missouri*. New York: F.P. Harper, 1903.

Clyman, James. *James Clyman, Frontiersman*. Edited by Charles L. Camp. Portland, Or.: Champoeg Press, 1960.

Colton, Walter. *Three Years in California*. New York: A.S. Barnes & Co., 1851.

Davis, William Heath. *Seventy-five Years in California*. San Francisco: John Howell Books, 1929.

Davis, William Heath. *Sixty Years in California*. San Francisco: A.J. Leary, 1889.

Delgado, James P. *Witness to Empire: The Life of Antonio Maria Sunol*. San Jose, Calif.: Sourisseau Academy for California State and Local History, 1977.

DeMers, Donald O., Jr. and Ann M. Whitesell. *Santa Clara Valley: Images of the Past*. San Jose, Calif.: San Jose Historical Museum Association, 1977.

De Voto, Bernard. *The Year of Decision, 1846*. Boston: Houghton Mifflin, 1961.

Dillon, Richard. *Captain John Sutter: Sacramento Valley's Sainted Sinner*. Santa Cruz, Calif.: Western Tanager Press, 1989.

Dillon, Richard. *Fool's Gold; the Decline and Fall of Captain John Sutter of California*. New York: Coward-McCann, Inc. 1967.

Donovan, Frank Robert. *River Boats of America*. New York: Thomas Y. Crowell, 1966.

Doten, Alfred. *The Journals of Alfred Doten. 3 vols*. Edited by Walter Van Tilburg Clark Reno: University of Nevada Press, 1973.

Dowling, Patrick J. *California, the Irish Dream*. San Francisco: Golden Gate Publishers, 1988.

Dowling, Patrick J. *Irish Californians*. San Francisco: Scottwall Associates, 1998.

Drago, Harry Sinclair. *The Steamboaters, from the Early Side-Wheelers to the Big Packets*. New York: Dodd, Mead and Co., 1967.

Egan, Ferol. *Fremont, Explorer for a Restless Nation*. Reno: University of Nevada Press, 1985.

Evans, E. Estyn. *Irish Heritage; the Landscape, the People and Their Work*. Dundalk: Dundalgan Press, 1942.

Fallon, Thomas. *California Cavalier: The Journal of Thomas Fallon*. Edited by Thomas McEnery. San Jose, Calif.: Inishfallen Enterprises, 1978.

Farquhar, Francis P. *History of the Sierra Nevada*. Berkeley: University of California Press, 1965.

Fava, Florence M. Los Altos Hills: *The Colorful Story*. Woodside, Calif.: Gilbert Richards Publications, 1976.

Fink, Augusta. *Monterey, the Presence of the Past*. San Francisco: Chronicle Books, 1972.

Fisher, Anne B. *The Salinas, Upside-Down River*. New York: Farrar and Rinehart, Inc., 1945.

Foote, Horace S. *Pen Pictures from the Garden of the World, or Santa Clara County, California*. Chicago: Lewis Publishing Co., 1888.

Fox, Frances L. *Luis Maria Peralta and His Adobe*. Sketches by Ralph Rambo. San Jose, Calif.: Smith-McKay Printing, 1975.

Graydon, Charles K. *Trail of the First Wagons over the Sierra Nevada*. St. Louis, Mo: Patrice Press, 1986.

Gudde, Erwin G. *California Place Names*. Berkeley: University of California Press, 1949.

Guinn, J.M. *History of the State of California and Biographical Record of Coast Counties*. Chicago: Chapman Publishing Co., 1904.

Hall, Frederic. *History of San Jose and Surroundings: with Biographical Sketches of Early Settlers*. San Francisco: A.L. Bancroft and Company, 1871.

Hammer, Jacob. *This Emigrating Company: The 1844 Oregon Trail Journal of Jacob Hammer*. Spokane, Wash.: Arthur H. Clark Co., 1990.

Hammond, George P. *The Weber Era in Stockton History*. Berkeley: Friends of the Bancroft Library, 1982.

Harlow, Neal. *California Conquered*. Berkeley: University of California Press, 1989.

Hartmann, Ilka Stoffregen. *The Youth of Charles M. Weber, Founder of Stockton*. Stockton, Calif.: University of the Pacific, 1979.

History of Santa Clara County, California. Introduction by J.P. Munro-Fraser. San Francisco: Alley Bowen & Co., 1881.

Hittell, Theodore H. *History of California. 4 vols.* San Francisco: 1885-97.

Holliday, J.S. *The World Rushed In*. New York: Simon and Schuster, 1981.

Hom, Gloria Sun. *Chinese Argonauts*. Los Altos Hills, Calif.: Foothill Community College, 1971.

Hopkins, Sarah Winnemucca. *Life among the Piutes: Their Wrongs and Claims*. Boston: 1883.

Hruby, Daniel D. *Mines to Medicine*. San Jose, Calif.: O'Connor Hospital, 1965.

Hunt, Rockwell D. *Personal Sketches of California Pioneers I Have Known*. Stockton, Calif., University of the Pacific, 1962.

Ignoffo, Mary Jo. Sunnyvale: *From the City of Destiny to the Heart of Silicon Valley*. Cupertino, Calif.: California History Center & Foundation, 1994.

The Irish: *A Treasury of Art and Literature*. Edited by Leslie Conron Carola. New York: Hugh Lauter Levin Associates, Inc., 1993.

Jacobson, Yvonne. *Passing Farms, Enduring Values: California's Santa Clara Valley*. Los Altos, Calif.: William Kaufmann in cooperation with the California History Center, De Anza College, Cupertino, Calif., 1984.

James, William F. and George H. McMurry. *History of San Jose, California*. San Jose, Calif.: A.H. Cawston, 1933.

Jones, William Carey. *"William Carey Jones Papers."* 1848-1918. Bancroft Library, University of California, Berkeley.

Kaler, Elizabeth. *Memories of Murphys*. 1956.

Keegan, Frank L. San Rafael: *Marin's Mission City*. Northridge, Calif.: Windsor Publications, Inc., 1987.

Kelly, Charles and Dale Morgan. *Old Greenwood*. Georgetown, Calif.: The Talisman Press, 1965.

Kennedy, Helen Weber. Interview. Stockton, Calif.: 1975-76.

King, Joseph A. *Winter of Entrapment: A New Look at the Donner Party*. Toronto: P.D. Meany, 1992.

Larkin, Thomas Oliver. *The Larkin Papers. 11 vols.* Edited by George P. Hammond. Berkeley: Published for the Bancroft Library by the University of California Press, 1951-68.

Lewis, Donovan. *Pioneers of California*. San Francisco: Scottwall Associates, 1993.

Lienhard, Heinrich. *A Pioneer at Sutter's Fort, 1846-1850: The Adventures of Heinrich Lienhard*. Translated, edited and annotated by Marguerite Wilbur. Los Angeles: The Califia Society, 1941.

Loomis, Patricia. *Milpitas: the Century of Little Cornfields, 1852-1952*. Cupertino, Calif., California History Center, 1986.

Loomis, Patricia. *Signposts II*. San Jose, Calif.: San Jose Historical Museum Association, 1985.

Lyman, Chester. *Around the Horn to the Sandwich Islands and California, 1845-1850*. Edited by F.J. Teggart, ed. New Haven, Conn.: Yale University Press, 1924.

Lyon, Mary Lou. *Captain Elisha Stephens, a True Pioneer and the Stephens-Murphy-Townsend Party of 1844*. Cupertino, Calif.: Lyon Historical Enterprises, 1995.

McGloin, John Bernard. *California's First Archbishop: The Life of Joseph Sadoc Alemany, 1814-1888*. New York: Herder and Herder, 1966.

McKevitt, Gerald. *The University of Santa Clara: A History, 1851-1977*. Stanford, Calif., Stanford University Press, 1979.

McKittrick, Myrtle M. *Vallejo, Son of California*. Portland, Or.: Binfords and Mort, 1944.

McNamee, Mary Dominica. *Light in the Valley; the Story of California's College of Notre Dame*. Berkeley: Howell-North Books, 1967.

Marmion, Anthony. *The Ancient and Modern History of the Maritime Ports of Ireland*. London: Printed for the author, 1858.

Mars, Amaury. *Reminiscences of Santa Clara Valley and San Jose*. San Francisco: Printed by Mysell-Rollins, 1901.

Morgan, Dale L. *The Humboldt: Highroad of the West*. Lincoln: University of Nebraska Press, 1985.

Nicholson, Loren. Rails across the Ranchos. Fresno, Calif.: Valley Publishers, 1980.

Noyes, Leonard W. "Flush Times in the Diggins, or the Lights and Shades of Murphys Camp in the Fifties, Being the Letters and Reminiscences of Captain Leonard W. Noyes." Edited and with an introduction and explanatory notes by Carl I. Wheat. Manuscript and letters, 1849-1858, Special Collections, California State University, Northridge.

O'Brien, Maire and Conner Cruise O'Brien. *A Concise History of Ireland*. London: Thames and Hudson, 1972.

Parkman, Francis. *The California and Oregon Trail*. Alexandria, Va.: Time-Life Books, 1983.

Payne, Stephen M. *A Howling Wilderness*. Los Gatos, Calif.: Loma Prieta Publishing Co., 1978.

Quackenbush, Margery, ed. *County Chronicles*. Los Altos Hills, Calif.: Foothill Community College District, 1972.

Rambo, Ralph. *Pioneer Blue Book of the Old Santa Clara Valley*. San Jose, Calif.: Rosicrucian Press, 1973.

Regnery, Dorothy F. *The Battle of Santa Clara, January 2, 1847*. San Jose, Calif.: Smith and McKay Printing Co., 1978.

Regnery, Dorothy F. *An Enduring Heritage: Historic Buildings of the San Francisco Peninsula*. Stanford, Calif.: Stanford University Press, 1976.

Rice, Bertha M. *The Women of Our Valley*. Vol. 1. San Jose: the author, 1955.

Robinson, W.W. *Land in California*. Berkeley: University of California Press, 1948.

Ryan, William Redmond. *Personal Adventures in Upper and Lower California, in 1848-9*. New York: Arno Press, 1973.

Sawyer, Eugene Taylor. *History of Santa Clara County, California*. Los Angeles: Historical Record Company, 1922.

Schallenberger, Moses. *The Opening of the California Trail*. Berkeley: University of California Press, 1953.

Schmidt, Earl F. *Who Were the Murphys: California's Irish First Family*. Second edition, revised and enlarged. Murphys, Calif.: Mooney Flat Ventures, 1992.

Shebl, James. Weber!: *The American Adventure of Captain Charles M. Weber*. Lodi, Calif.: San Joaquin Historical Society, 1993.

de Smet, Pierre-Jean. *Letters and a Sketch of a Trip to the Rocky Mountains*. Philadelphia: W. Fithian, 1843.

Starr, Kevin. *Americans and the California Dream, 1850-1915*. New York: Oxford University Press, 1973.

Stewart, George R. *The California Trail*. Lincoln: University of Nebraska, 1983.

Stone. Irving. *Men to Match My Mountains; the Opening of the Far West, 1840-1900*. Garden City, NY: Doubleday, 1956.

Sullivan, Gabrielle. *Martin Murphy, Jr., California Pioneer, 1844-1884*. Stockton, Calif., Pacific Center for Western Historical Studies, University of the Pacific, 1974.

Taylor, Bayard. *Eldorado, or, Adventures in the Path of Empire*. New York: Alfred A. Knopf, 1949.

Thompson & West. *Historical Atlas Map of Santa Clara County*. San Francisco, Calif.: 1876. Printed by Smith & McKay Printing Co., San Jose, Calif.: 1973.

Tinkham, George H. *History of San Joaquin County, California*. Los Angeles: Historic Record Co., 1923.

Tinkham, George H. *A History of Stockton from Its Organization up to the Present Time*. San Francisco: W.M. Hinton and Co., 1880.

A Treasury of the Sierra Nevada. Edited by Robert Leonard Reid. Berkeley: Wilderness Press, 1983.

Under the Shadow of El Toro. Compiled by Joyce Hunter. Morgan Hill, Calif.: M. Stinnett and J.L. Stinnett, 1978.

Uris, Leon and Jill Uris. *Ireland, a Terrible Beauty*. Garden City, New York: Doubleday, 1975.

Waggoner, Madeline Sadler. *The Long Haul West; the Great Canal Era, 1817-1850*. New York: Putnam, 1958.

Wallace, Katherine. *California through Five Centuries*. New York: AMSCO School Publication, Inc., 1974.

Walsh, Henry L. *Hallowed Were the Gold Dust Trails*. Santa Clara, Calif.: University of Santa Clara Press, 1946.

Walsh, James P. *The San Francisco Irish, 1850-1976*. San Francisco: The Society, 1978.

We Were '49rs: Chilean Accounts of the California Gold Rush. Translated and edited by Edwin A. Beilharz and Carlos U. Lopez. Pasadena, Calif.: Ward Ritchie Press, 1976.

Wood, R. Coke and Leonard Covello. *Mother Lode Memories: A Pictorial History*. Fresno, Calif.: Valley Publishers, 1979.

Wood, R. Coke. *Murphys, Queen of the Sierra*. Angels Camp, Calif.: Calaveras Californian, 1948.

Wood, R. Coke. *Tales of Old Calaveras*. Angels Camp, Calif.: Calaveras Californian, 1949.

Woods, Daniel B. *Sixteen Months at the Gold Diggings*. New York: Arno Press, 1973.

Wyman, Beth. *Hiram Morgan Hill*. Morgan Hill, Calif.: By the author, 1983.

Young, John V. *Ghost Towns of the Santa Cruz Mountains*. Santa Cruz, Calif.: Paper Vision, 1979.

Zauner, Phyllis. *Those Spirited Women of the Early West: A Mini-History*. Sonoma, Calif.: Zanel Publications, 1989.

Zollinger, James Peter. *Sutter; the Man and His Empire*. New York: Oxford University Press, 1939.

Newspapers and periodicals-

Warren, E.W. "Re: John Murphy and His Indian Wife Pokela." *Daily Evening Transcript* (Boston) November 13, 15,1849

Gilroy Advocate, November 4, 1922

Mountain View Register, April-June 1894

Oakland Tribune, July 19, 1942

Salinas Californian, November 15, 1958

San Francisco Chronicle, April 26, 1970

San Francisco Chronicle, May 4, 1970

San Francisco Chronicle, June 3, 1970

San Jose Mercury, July 1, 1882

San Jose Mercury, January 7, 1951

Index